FOREST GOVERNANCE and SUSTAINABLE RESOURCE MANAGEMENT

FOREST GOVERNANCE and SUSTAINABLE RESOURCE MANAGEMENT

Irshad A. Khan

Los Angeles | London | New Delhi
Singapore | Washington DC | Melbourne

First published in 2019 by

SAGE Publications India Pvt Ltd
B1/I-1 Mohan Cooperative Industrial Area
Mathura Road, New Delhi 110 044, India
www.sagepub.in

SAGE Publications Inc
2455 Teller Road
Thousand Oaks, California 91320, USA

SAGE Publications Ltd
1 Oliver's Yard, 55 City Road
London EC1Y 1SP, United Kingdom

SAGE Publications Asia-Pacific Pte Ltd
18 Cross Street #10-10/11/12
China Square Central
Singapore 048423

Published by Vivek Mehra for SAGE Publications India Pvt Ltd, typeset in 10.5/13 pt Adobe Caslon Pro by Zaza Eunice, Hosur, Tamil Nadu, India.

Library of Congress Cataloging-in-Publication Data Available

ISBN: 978-93-532-8195-3 (HB)

SAGE Team: Abhijit Baroi, Sandhya Gola and Nishant Dhawan

*To
my wife Shaheena,
my daughter Shazia
and
my son Ahmer*

Thank you for choosing a SAGE product!
If you have any comment, observation or feedback,
I would like to personally hear from you.

Please write to me at **contactceo@sagepub.in**

Vivek Mehra, Managing Director and CEO, SAGE India.

Bulk Sales

SAGE India offers special discounts
for purchase of books in bulk.
We also make available special imprints
and excerpts from our books on demand.

For orders and enquiries, write to us at

Marketing Department
SAGE Publications India Pvt Ltd
B1/I-1, Mohan Cooperative Industrial Area
Mathura Road, Post Bag 7
New Delhi 110044, India

E-mail us at **marketing@sagepub.in**

Subscribe to our mailing list
Write to **marketing@sagepub.in**

This book is also available as an e-book.

Contents

Part V: Emerging Global Issues: Commitments and Challenges

Preface

Why India, despite having all the essential ingredients, is unable to make significant progress in scientific and technological innovations and is lagging behind developed countries appears to be an enigma. But the root causes are hidden in cultural complexities, human behaviour and lack of good governance. Though necessary policies, laws, rules and regulations, processes and procedures, institutional structures and other resources are available, there is a lack of innovation and creativity, and scientific and technological progress is not being made. An inertia has set in, which has become so deeply entrenched in the system that it appears unlikely to be broken in near future. There are good intentions and desires, but the corresponding actions are inadequate. There is a huge mismatch between desire and attainment.

Personal interests are held supreme by the people who manage the affairs of the state at all levels of governance. The Indian bureaucracy, which inherited the colonial mentality and added to it the indigenous feudal temperament, could have delivered much more for the country if it could have been sincere, committed, forthright and displayed professional integrity. But its members always focused on their self-interests in terms of postings, for which they found no alternative but to align with and serve the interests of self-serving power centres. The forest officers of India are also part of this system. The 'system' is so powerful and overwhelming, so deeply entrenched, that it acts ruthlessly against non-conformist elements and constructive dissent.

What Does Society Expect from the Forestry Sector?

This is an important question that must be addressed before any discourse on the forestry sector is taken up. Another issue is an understanding of what society perceives about forestry institutions, and that needs a very thorough and detailed analysis.

Broadly, society has many stakeholders as far as the forests are concerned. There are city dwellers who look to the forests for aesthetics, ecotourism, conservation, wildlife, water resources as well as timber for house construction and furniture. The industrialists and traders focus on the supplies of raw material and marketable products. There is a demand for clearing forests for mining of coal, ores, stones and other minerals. The rural communities that interface frequently with the forests or dwell inside them need sustained supplies of fuelwood, small timber, fodder and pasture land, and non-timber forest products (NTFP) for their subsistence and livelihood needs. There is another class of stakeholders who exert significant influence on the management of forests—this includes politicians, bureaucrats and the media.

There are now two additional demands on forests. One is carbon sequestration from the atmosphere and its storage to aid climate-change mitigation. The second is conservation of biodiversity. On the other hand, there is a huge pressure on the forests from all sides that is impossible to meet with the scarce forest lands and forest resources. Added to this are vested interests that play a role that is detrimental to forest conservation and regrowth. This complexity specific to the forest sector alone has been resulting in conflicts that have lately been exacerbated. The forestry personnel have not been able to deal with these conflicts and have reacted by acquiring a passive attitude over the decades. Many forest officers at different levels express pessimism and a sense of resignation.

- Why has this situation arisen?
- What circumstances caused this state of affairs?
- What have been the historical changes?

These and many other questions provide the basis for an objective study from within and outside the sector.

The problems that the forestry institutions face do not have technical solutions. A knowledge or mastery of forestry science alone will not help. However, modern skills and strategic approaches will be very useful to get out of the current state of affairs. The failure of the forestry institutions is the failure of Indian governance. There are policies and laws that are adequate to facilitate the management of forest resources on a sustainable basis. However, implementation of the policies and laws has been weak or ineffective. A revision of forest policies will not be of much help unless the causes of their non-implementation or ineffective implementation are reviewed and lessons learnt. An aspirational policy without concomitant resources in terms of finance, political and social support and institutional capacity will remain on paper only, as has been the experience with the last two national forest policies adopted.

It is also acknowledged that the lack of good governance is not confined only to the forestry sector in India, but applies to all sectors of the economy, including law enforcement, revenue management, public works, irrigation, education, health, municipal bodies, public utilities, energy and transportation sectors. One can easily find commonalities among the relevant institutions, namely the inadequate implementation of sectoral policies, rules and procedures across the governance landscape.

This book is expected to be useful to forestry academicians, researchers and advanced students in various universities, training institutions and management schools as well as to civil servants, government institutions and civil society both within and outside India. The book will help in deepening readers' understanding of the complexities of forest governance and its relationship with the continued depletion of forest resources.

Introduction

Of late—that is, for the past three decades—India's national focus has been on economic growth. The gross domestic product (GDP) has been growing; so also the average inflation. It is claimed that poverty has reduced, though 350 million people are still below the poverty line. Economic growth has also brought economic disparities, as the rich are becoming richer. The future of forestry and poverty are closely related. If poverty is reduced, people have alternative sources of domestic energy and shift to stall-feeding of their livestock, then dependence on the forests and conflicts over their resources and land use will be reduced. Thus, economic growth has to be inclusive, with equity for all. The poor have to be provided opportunities which are otherwise seized only by those who are better placed in our society. This affects forests, as poverty is one major cause of environmental degradation.

Revenue from Forests and Contribution to GDP

Revenue from the forest sector has declined in real term. Forest productivity and output have declined. More timber is imported than produced domestically. Per land unit, the growing stock is more outside the forests than inside. (Please see Chapter 4.)

Alliances and Conflicts

Do forestry institutions have allies outside? We have attempted to identify various stakeholders. Do they share the views of the foresters relating to sustainable management of natural resources, conservation, the role of forests in climate-change mitigation, the conservation of watersheds for a source of clean water, restoration of degraded forests, and so on? Maybe some enlightened intellectuals who are urban may appreciate it, but the rest of society would likely prefer coal, iron, timber, roads, transmission lines, hydel stations, railway lines, irrigation canals and building construction at the cost of the forests. Sincere and committed forestry personnel face frequent or perpetual frustration over the public's contempt for forestry values. Most stakeholders are dismissive of the concern that is expressed by foresters. Forestry institutions are regarded as hindrances to the development projects, on one hand, and threats to forest-interfaced communities as the guardians of forest resources against indiscriminate plunder (which they fear would happen if local communities are allowed laissez-faire), on the other hand.

Forestry in India: Perceptions and Realities

That inextricable conflicts of interest are present in the forestry sector is a reality, and solutions continue to elude us. Also, how many foresters work together collectively to address the issues and make a serious attempt to manage and resolve conflicts? Are the foresters able to convince the stakeholders of the need to sustainably manage forests? Cross-sectoral policies and priorities also occasion serious conflicts. For example, the tribal welfare lobby or departments would advocate, like some pro-tribal NGO activists do, handing over of all forests in tribal areas to the Adivasi communities or at least allow them the freedom to use the forests and clear land for agriculture. The animal husbandry or livestock sector promotes goats and sheep, without assessing need for or making available fodder and pastures. The mining sector demands forest clearance for mining activities. The roads and highways sector wants land for new road construction and widening of existing ones through forests and protected areas (PAs). Politicians support encroachment on forest lands and its regularization. The irrigation sector wants forest clearance for water reservoirs and canals. These other sectors thus have conflicting interests, which constitute the very opposite of the task of the forestry personnel. This conflict also adopts the guise of one between human welfare and the protection of elephants and tigers and trees. To many, therefore, the forest department appears as an enemy, hurdle or inconvenience.

For how long and to what extent may we allow the loss or degradation of forests? Society does not have the answer to this. It expects that forests will be conserved and, at the same time, society's needs from the forests will be satisfied.

The evolution of forestry in India has a chequered history. For more than 100 years, the main job of the state forest departments (SFDs) was to cut wood through contractors. Contracts (called 'concessions' in the Western world) were auctioned. In the 1970s, forest corporations started appearing, the first one being the Forest Development Corporation of Maharashtra (FDCM). In the 1980s, there came the wave of afforestation and social forestry, which waned in the mid-1990s. In the 1990s, instead participatory forest management (PFM) or joint forest management (JFM) was promoted, the scaling up of this being the main thrust of forestry policy. This trend is also now showing signs of waning. Judicial activism, since 1995, influenced forest management and also deterred the indiscriminate cutting of forests, which was resulting in extensive destruction. We do not know what is next. All these developments were, strangely, accompanied by a stagnation in the forestry sector and an inertia in the forestry institutions. The current phase is still characterized by inertia in all aspects of activities. This inertia is reflected in a lack of initiative in research and innovation, both in forest management and forestry education.

Global Developments

All over the world, steps have been taken and are being taken to address the need for sustainable forest management and an integrated management of forest resources. There are many stakeholders all over the world who are calling today for a paradigm shift in forestry. It is being argued that old practices, based on the priority of timber above all other functions and values of forests, are obsolete and that a new forest management paradigm is needed. Such arguments are based on the perception that forest managers have remained obsessed for the past 200 years with a paradigm where the production and harvesting of timber have been dominant. It is a fact the forestry practices have recently been undergoing very deep and rapid changes. However, this new paradigm does not mean the complete end of 'traditional forestry' as we know it today. We only have to change our approach to forest management so that it is science-based and ecosystem-centric.

Multifunctional, multiple-use and holistic approaches will increasingly become the basis for managing forest resources. The change from classical, sustained-yield management of a few commercially important species to a focus on the conservation and sustainable use of forest ecosystems is transforming some of the basic principles of forest management. The trend, however, is that the demand for wood and wood products and for ecosystem services is rising gradually and the productive forest area is declining, particularly in tropical countries, due to deforestation, degradation and use of forest land for non-forest purposes. These trends are resulting in increased pressure for wood production from forests as well as for non-timber forest products (NTFPs), biodiversity and other ecosystem services.

International Action

Forestry personnel historically thought that they were managing forests on a sustained-yield basis and believed that this was the way that forests would be managed in the future too. The myth that foresters are omniscient has been shattered long ago. Globally, forestry is now being guided by the following developments:

The Rio Declaration (1992)

The Rio Declaration on Environment and Development included 27 principles defining the rights and responsibilities of nations as they pursue human development and well-being. One of the principles was that in order to achieve sustainable development, environmental protection should constitute an integral part of the development process and could not be considered in isolation from it (Principle 4). Another important principle was that states should cooperate in a spirit of global partnership to conserve, protect and restore the health and integrity of the Earth's ecosystems (Principle 7).

Agenda 21

Also adopted in Rio, Agenda 21 reflected a global consensus and commitment at the highest political level to make development socially, economically and environmentally sustainable. A non-binding but authoritative Statement of Principles for the Sustainable Management of Forests was also adopted, to guide the management, conservation and sustainable development of all types of forests.

United Nations Framework Convention on Climate Change

The UNFCCC aims to stabilize the levels of greenhouse gases (GHGs) in the atmosphere at concentrations that will not dangerously upset the global climate system. It was also adopted in Rio in 1992, based on which the Kyoto Protocol was adopted in 1997 by most countries.

Convention on Biological Diversity

More than 150 countries at the Rio Summit signed the Convention on Biological Diversity in 1992. It became effective on 29 December 1993.

Ecosystem management

The ecosystem approach to forest management is the new paradigm. There were many initiatives to develop criteria and indicators for sustainable forest management. However, the application of such criteria and indicators to forest management and forestry policy is still far away. There are many hurdles in understanding the concept of sustainable forest management and therefore to implement it. It would be advisable to move to a new, broader paradigm, which would enable us to improve forest management to address the ecological, economic and social needs of both present and future generations. This new paradigm is ecosystem-based management.

Ecosystem-based forest management evolved in the United States of America (USA) in response to court orders curtailing the production of timber by half in many states (within the range of the spotted owl). At the Rio Conference, the USA declared that it was going to adopt an ecosystem approach to forest management. The President of the USA, Bill Clinton, convened a full-day conference in Portland, Oregon, on 2 April 1993, which decided to set up a working group, the Forest Ecosystem Management Assessment Team (FEMAT). That was beginning of forest ecosystem management.

REDD Plus

Another recent development is the adoption of the Reduction of Emissions from Deforestation and Degradation of Forests (REDD plus) approach to mitigate climate change in tropical countries. It focused on reduction of emissions from deforestation and forest degradation; sustainable management of forests, conservation of carbon stocks and enhancement of carbon stocks. It is also a part of the Paris Agreement adopted by all countries party to the UNFCCC in December 2015. REDD plus is to be implemented by developing countries which are experiencing large-scale deforestation and forest degradation. Please see Chapter 11 for details.

India and the world

India will soon be required to integrate its forestry practices with global forest policies and expectations. India's economy, culture and technology are already rapidly integrating with global standards. So it is a matter of time before Indian forestry comes out of its isolation and integrates with global environmental efforts. Climate change is already a global issue and the forests are expected to play a significant role. The global changes would require additional and changed roles of forestry personnel in the coming years. As new forest policies are conceptualized as practical instruments that are feasible and imple-mentable, new responsibilities will devolve on foresters.

Evolution of Forestry in India

There have been many changes in forestry in India since 1950. Though India's forests are not meeting its full demand for timber and the country is importing huge quantities of timber and wood pulp, there is as such no crisis over the supply of wood. Thus the supply of timber worldwide is not showing any perceptible crisis. Only if timber import is affected will there be a serious crisis.

India's forestry management and policy trajectory took a somewhat new direction in the 1970s and the 1980s, partly under its own compulsions and partly due to an outcry against forest and wildlife depletion worldwide. These concerns were reflected in the Wildlife (Protection) Act of 1972 and the Forest (Conservation) Act (FCA) of 1980, with an increased emphasis on afforestation and social forestry. The other policy directions included phasing out of timber contractors and their replacement by state-run forest corporations. The new changes necessitated new skills among forestry personnel, such as forest logging, transportation of wood, forestry extension, engaging people's participation, agroforestry, and so on. Interestingly, the FCA brought new and unending conflicts by its restriction of arbitrary and forceful conversion of forest lands to non-forest uses by the states, creating an atmosphere of resentment against forestry institutions, which came to be seen as a hindrance to development and the well-being of the people.

Policy Implementation

The National Forest Policy of 1952 became irrelevant and deliberations started on its revision in the 1970s. The new policy was adopted in 1988. The FCA of 1980 was amended in 1988 and was made more stringent. Despite these, deforestation and forest degradation continued. The massive afforestation drive of the 1980s was followed by another management paradigm that demanded community participation in the protection of forests in lieu of a share in the forest produce. Later on, this was popularly called 'joint forest management' (JFM), which according to many activist NGOs was 'neither joint nor forest management'.

Impact of JFM

The actual efficacy and impact of JFM on forest growth and productivity are yet to be fully understood. The question that is still being asked is whether JFM was willingly adopted by forestry institutions or whether it was thrust upon them under the guise of finance for externally aided programmes or development schemes. Whatever the reason for its initiation, promotion and expansion, forestry personnel were under tremendous pressure to change their mindsets and learn the art of participatory management. Reorientation and training programmes involving forestry personnel, local communities and civil society were initiated and are still going on.

Judicial activism

Judicial activism came as a salvaging act. It has had a profound impact on the way forests were managed and came about as a result of the failure of the executive to implement its own policies, allowing rampant destruction of forests in North-East India and elsewhere. It has had many positive results—one has been a halt to the indiscriminate and heavy exploitation of forests by a combination of vested interests. The working-plan system was restored, sawmill activities regulated, felling of green trees banned in many states, net present value (NPV) and compensatory afforestation funds (a kind of payment for environmental services) institutionalized, and so on. However, the executive has so far failed to fully respond to the judicial verdicts to renew and ensure sustainable management of forests through policy, legal and financial inputs.

Challenges

The challenges for the forest sector and therefore for forestry institutions are still growing. There are expectations to increase the production of wood and other forest products as well as enhance a host of ecosystem services. At the same time, the challenge is harmonization of the competing demands on forests with the local needs and livelihood issues of forest-dependent communities. Political support for forestry is not significant and the environment is not a priority. The priority today is fast economic growth, measured in terms of GDP. This makes it unavoidable to rethink and revitalize the role and significant contributions of forestry in national economic development.

Changing roles

Changes in forestry practices are inevitably accompanied by changes in the role of foresters. The forestry personnel of the present and of the future will be dealing with new expectations, responsibilities, accountabilities and also challenges. They are faced with a serious challenge to overcome public apathy and lack of social awareness and support for forest conservation and sustainable use. Foresters have historically exhibited a tendency to view the world from their point of view, which limits their ability to respond to different points of view, both rational and irrational.

Governance deficit

There are serious governance issues in forestry, attributable partly to external forces that constrain foresters' effectiveness and institutional proactivity, which tend to make them passive participants. The siege mentality among foresters has been prevalent for the last four decades or more. They have not been able to come up with a strategy to counter the unjustified criticism and blame games and give up the sense of remorse by developing a greater sense of self-confidence and self-esteem. This can be achieved only when they practice good governance. They also need to be empowered to play a crucial role in the management of natural resources and to provide environmental services for forest ecosystems. Governance reforms in general and in the forestry sector in particular will provide an enabling environment to manage forest ecosystems on a sustainable basis.

Structure of the Discussion

This book is divided into five parts and 11 chapters. The preface and introduction will help readers to appreciate the background and focus of the contents. Each chapter is self-contained, however, and readers may choose what interests them.

Chapter 1 explores how and under what circumstances the British colonial regime in India acquired forested land and brought European practices to manage the tropical forests in this country, initially with the help of German and French forestry professionals. After Independence (1947), the colonial management practices continued, with an increased focus on the maximization of revenue generation for the state governments. Later, the environmental movement the world over and also in India triggered conservation efforts, resulting in new policy initiatives and programmes.

Chapter 2 discusses the evolution of forest policy from the British period to independent India, with a focus on the failure of the National Forest Policy of 1952 and the consequent adoption of a new policy in 1988. A detailed analysis of the implementation of the current national policy of 1988 and its outcomes are given in a matrix in the Appendix at the end of the volume. Chapter 2 also includes a discussion on new initiatives and approaches since 1950, their outcomes and sustainability, as well as missed opportunities to revisit and rewrite realistic and achievable policies.

Chapter 3 analyses the influence of external donor agencies that provided significant financial assistance for forest development in India through state-wide projects since 1979, and their impact. The World Bank, in particular, used its leverage to push through limited but significant reforms in the public forestry institutions. This chapter also includes a brief account of the World Bank's dilemma over continuing to support the forest sector, due to its own internal wrangling and the senior management's risk aversion.

Chapter 4 examines the state of forest resources in India on the basis of available information and actual conditions observed on the ground in limited areas. Important questions that are posed include why the efforts of massive afforestation and reforestation have not been successful in expanding forest cover and productivity, and why further degradation could not be arrested.

Chapter 5 provides a discussion on the management of state-owned forests in India. The condition of forests clearly reflects the impact of past management policies and practices. The management style and field operations neither could provide adequate support to achieve the policy objectives, nor could they reverse the trends of forest ecosystem degradation. There is an emphasis on the need to change the way forests are managed, also necessitating good governance.

Chapter 6 covers the social forestry movement initiated by the government, with funds from mainly multilateral and bilateral international aid agencies and partly from its own domestic sources as well. The programme saw success in terms of increased biomass production, agroforestry expansion and awareness among people. However, with the gradual withdrawal of donor support, the programme could not be sustained at the level it had achieved during the 1980s.

Chapter 7 examines the policy and outcomes of community participation, mainly in the protection of public (state) forests through JFM, officially adopted in the 1990s. This involved the setting up of community-based institutions with the aim of promoting equity and democratic decentralization among forest-fringe rural areas, with an arrangement of sharing forest produce or income or both from the forest area assigned for joint or participatory management.

Related to community participation is the focus on poverty alleviation, which has been described in Chapter 8. Though forests play a role in the livelihoods of forest-dependent communities, there are limitations. Poverty reduction from forest management has been mainly pushed by the external donor agencies, and was later adopted as one of the outcomes of JFM.

In Chapter 9, the main forest laws and their enforcement have been discussed. An important impact on forest management has been made by judicial interventions since the mid-1990s. The Supreme Court of India effectively put a halt to the ongoing destructive timber-harvesting practices. The controversy surrounding the Forest Rights Act of 2006 has been briefly touched upon.

Chapter 10 is the key chapter of this book, which builds upon all previous chapters. In it, the specific attributes of forest governance in India have been analysed. Though the quality of forest governance varies from state to state, an attempt has been made to objectively assess the status of governance in the overall forest sector in the country, with a focus on accountability and transparency.

Chapter 11 provides a discussion of the role of forests in global climate-change mitigation, implying that the whole world is tending to become a stakeholder in all forests, irrespective of the political boundaries. Forests are both sources and sinks for GHGs. India is a signatory to the Paris Agreement and has committed through its intended nationally determined contribution (INDC) pledge to apply REDD+ principles and to create additional carbon sinks by afforestation and reforestation. It is to be seen how this global commitment is fulfilled by 2030.

Finally, an Appendix comprising a detailed matrix analysing the implementation of the National Forest Policy and its outcomes or achievements and its constraints is given at the end. It can be thought of as a supplement to Chapter 2.

This book is the first of its kind in India to explore the relationship between good governance, trends and other intricacies involved in the multiple-use and multiple-stakeholder natural resource that forests represent, a resource linked to human survival on our planet. Other publications on forest policies may provide prejudiced interpretations focused on perhaps a specific stakeholder group, or simply a narrative of facts or information. There are no publications in India or abroad at this time that present a comprehensive analysis. However, the author refrains from providing solutions or recommendations, but attempts to offer an unbiased and objective policy analysis. The book examines policies that could have highly desirable outcomes if implemented as per aspirations.

In summary, whereas the British colonial administration had a purely extractive management agenda for valuable timber species, which motivated it to acquire this common property resource (then so-called 'wastelands'), independent India's forest policies did not help in the prevention of indiscriminate exploitation of forest land and biomass either. The challenge has been inconsistency between policy and management goals on the one hand and finance, infrastructure and capacity of institutions to implement these, on the other. As a result, new initiatives, approaches and programmes that did contribute in improving the situation still achieved only limited success and could not be sustained.

Part I

Forest Management and Policy Evolution

CHAPTER 1

Organized Forestry
Inception

The history of forestry in India is a history of conflicts and competing demand on resources. Systematic forest management started in the middle of the nineteenth century with the process of settlement and reservation of forests. India's colonial rulers brought German and French foresters to India, thereby introducing Franco-German forest management practices in this country. Dietrich Brandis, the first Inspector General of Forests, was a German and the second, Sir Ribbentrop, was French. The early foresters in India were trained in Germany and France.

The main aim of the British was to gain control over the forest resources to extract timber for their shipbuilding industry in Britain and for railway sleepers to lay a railway network in India for rapid movement of troops to have control over the country. The Forest Policy of 1894 gave agriculture priority over forestry. The Forest Policy of 1952 cautioned against the diversion of forestland to agricultural use and aimed at having one third of land under forest cover. However, the forests have depleted more rapidly since then. The current policy enunciated in 1988 had the same cherished aspiration of bringing under forest cover one third of the country's total geographical area, the attainment of which appears to be a remote possibility or a distant dream today.

The history of India's forest management is very fascinating as it had linkages with European forestry, from where it borrowed most of its practices and principles. Before organized forestry was initiated in India by the British colonial regime, Europe was already experimenting with forest management systems that mainly evolved in Central Europe. The German and French forestry scientists responded to society's demand to make forestry more profitable for landowners and to also ensure fast and attractive returns from the harvest. Foresters were concerned only with trees and how those trees yielded the product which was to meet the market demand. As the native species were mostly slow-growing hardwoods and needed long rotations, attention shifted to short-rotation softwood species. Conifer species such as fir and spruce were fast-growing. This led the practitioners of forestry to clear-cut the land of native species and plant a monoculture of conifers.

In Europe, vast stretches of lands had even been rendered degraded due to heavy cattle grazing and fuelwood harvesting for domestic energy. Both publicly and privately owned land had become considerably unproductive. In the late eighteenth and early nineteenth centuries, the European landscape underwent changes. Artificially grown pine forests dominated land use. The product was in demand and huge exports were made to Western Europe. Even where forests were managed through natural regeneration (e.g., the Normandy province in France), major commercial species such as oak and beech were preferred, and pure crops of these two tree species were encouraged through manipulation of the canopy, successive and heavy regeneration cuttings, and clearance of the so-called undesirable species.

When the British imperial government brought Dietrich Brandis from Burma (now Myanmar), where he was Forest Superintendent, to India as the first Inspector General of Forests, heavy exploitation of forests was causing concern to the colonial administration. The Madras Presidency and Travancore had already witnessed heavy logging by vested interests, and conservators of forests had been appointed there even before Brandis came. Brandis, who was a professor of botany and forestry in Bavaria (Germany) before he was brought by the British to this part of the world, brought with him forestry practices from Central Europe. The colonial British forest policy was well thought out, with a focus on the exploitation of India's rich natural resources to the maximum extent possible to earn revenue for the Empire. The concern for conservancy was meant to ensure this revenue would be available in perpetuity on a sustained basis.

Forest Exploitation to Serve Colonial Interests

The British planned to lay a railway network for facilitating quick movement of troops from one part of India to the other part to quell rebellions against the Empire. This involved cutting down the best wood for railway sleepers. Sal (*Shorea robusta*) and teak (*Tectona grandis*) were heavily extracted to meet the demand for railway sleepers. To lay down 2 km of railway line, 900 sleepers were needed. The railway network that was 32 km long in 1853 became 51,650 km in 1910. During various deliberations of the colonial administration in India and in Britain about India's forests, the agreement was to continue this exploitation for the maximization of revenue.

In Europe, alongside exploitation, investment was also made in regrowing trees or reforestation in harvested areas. However, in India, the British regime was disinterested in investing in or incurring expenditure for the regeneration of plantations in harvested areas. According to them, it would have reduced the net profit from harvesting. The approach was to introduce management systems that did not involve replanting but relied on elusive natural regeneration. The shelterwood system was to earmark areas and cut down everything except a few trees of commercially valuable species whose seed could regenerate the area. This involved total removal of those tree species which were not fetching a good price or had no market demand. The species that were favoured in the Himalayas were *Cedrus deodara* (deodar), *Pinus wallichiana* (blue pine), *Pinus roxburghii* (*chir* pine), fir and spruce. In the tropical parts of the country, sal, teak, rosewood and sandalwood were favoured. The other management system adopted was selection of individual trees or a group of commercially valuable trees for cutting, and leaving less important and small-sized trees of valuable species. The selective tree cutting too was guided by financial considerations. The other possible systems were clear-cutting and plantations. This required a huge amount of funds, which was not forthcoming from the treasury. There were nevertheless plantations, and many of them were very successful. Large-scale teak plantations were raised only later. In the twentieth century, the coppice system was adopted in the cases of teak and sal. This was an economical method to regrow trees on harvested land by allowing the stumps to grow shoots, and by tending the best-growing, most vigorous shoots, the logged-over areas were regenerated.

These types of forest management systems were introduced not only in the British-administered provinces but also in the princely states. Brandis worked to consolidate India's forest estates by settlement and demarcation. An efficient planning system was adopted for producing working plans, which is proudly referred to as the first planning system of any kind in India. These working plans were produced for each forest division in the country. They contained information about the forests and an estimation of growing stock, with a prescription of how much should be cut down annually and where. The principle of sustained yield was also borrowed from European forestry management principles.

In 1935, forestry as a subject was transferred to the jurisdiction of the provinces; in 1938, the Imperial Forest Service was discontinued and state forest services, or the Superior Forest Service, established. Forestry research and education continued to be centrally controlled from Dehradun's Forest Research Institute. Forest laws were enacted and enforced to protect forests. In the process of forest settlement and reservation of forests, local community rights were recorded, but not in the case of nomadic tribal communities that were dependent on hunting and food gathering and were not settled agriculturists. Many tribal activists denounce today the fact that in the process, the forests were stolen from the tribal communities and forest dwellers, which deprived them of their livelihoods and way of life. Forest conservancy meant that people had to be kept away from the forests and that there could be no exploitation without permission. Illegal cutting of trees and encroachment or illegal occupation of forest land would be severally punished. However, the traditional rights of local people were recognized and they were given concessions for the collection of forest products for their bona fide domestic needs. Cattle trespasses were also restricted and any intrusion by cattle without a grazing permit became an offence. Thus, what used to be no man's lands, wastelands or common property became resources acquired by the state or the ruling regime, be it the British colonial administration or the feudal administration of kings and princes.

In 1884, the colonial forest policy recommended clearance of forests to make way for agriculture to feed India's ever-growing population. Around villages, sizeable chunks of forests were left out from reservations and government control for use by the local people. These lands under the revenue departments' control were mostly illegally grabbed or officially distributed for agriculture, resulting in the disappearance of buffers.

Post-Independence Colonial Legacy

When India gained independence at midnight on 15 August 1947 and made a 'tryst with destiny', its forests had already undergone substantial depletion. When the feudal system was abolished in 1948, the feudal lords had ensured that by then the forests were cleared of trees. These deforested lands and thousands of hectares of degraded and denuded revenue lands were handed over by the state revenue departments to the state forest departments (SFDs) for management. The merger of the princely states into the Indian Union also brought various kinds of land tenure and forest management traditions and issues. In many states, there were fragments of land where tenure was not clear and the forest settlement process was not complete, giving rise to potential disputes. The land tenure dispute has still not been resolved in many areas.

After India became a republic and adopted a constitution, governance issues came to the forefront. The colonial legacy continued as a matter of administrative expediency. There was, even before 1947, a growing concern over forest depletion and therefore a national forest policy was prepared and adopted in 1952. It laid emphasis on discouraging clearance of forests for agriculture, in a departure from the previous policy of 1894, though this policy was also aimed at the exploitation of forests and sought to carry on the colonial legacy. The provincial forestry establishments continued to employ colonial forest management practices that had as their main aim the harvesting of forests to maximize revenue for the state exchequer, and then to protect the forests from illegal activities. The expectation of sustained yield became progressively increasing yield of forest products and thereby revenue. All working plans prepared or revised were guided by this principle of the generation of the maximum possible revenue. The annual revenue targets from the forestry sector were set by the state finance departments without due regard to the growth and capacity of the forests to provide sustained yield in perpetuity.

Foresters in India, like their European counterparts, were now obsessed with the idea of and aspirations for a 'normal forest', which was more a matter of fantasy and fiction than reality. All working plans invariably had an overarching objective of management to attain a 'normal forest' that would have an ideal growing stock, ideal increments and ideal age gradation. The concept involved forests of different ages being classed by growing stock and growth that should be transformed in such a way that the yield of timber could be sustained in perpetuity. This objective failed miserably and no one in the forestry sector now talks about a 'normal forest'. *However, it took more than a century for us to stop chasing this mirage.* Forestry practices in Europe were heavily copied from agriculture, and forests were even called 'crops'. But the logistics of an annual crop, which most agricultural crops are, were difficult to replicate through trees. As forestry was not an established science, the only lessons the forestry practitioners learned were from experience in agriculture, that is, to harvest and regrow.

The forest administration carried on with the colonial tradition and, compared to earlier, it displayed even greater enthusiasm for higher revenue receipts at all costs. There was even a kind of competition among forest officers for revenue mobilization from their respective jurisdictions. They were oblivious to the question of what would happen after the heavy exploitation, so long as there still remained other forest areas to harvest. Forest logging was so heavy and reckless that it had a semblance to mining. The net consequence of this practice was severe depletion of forest resources. There were serious allegations that there was an unholy nexus among contractors, politicians and forest officers—who were all out to make a quick buck by indulging in unethical practices. These allegations were not unfounded and forest-dominated states had huge corruption scandals.

Phases

There have been distinct but partly overlapping phases in India's forest policy and management evolution. These can be summarized as shown in Table 1.1.

New Developments

In protest against the reckless destruction of forests, a nascent environmental movement created significant public awareness against tree felling. The Chipko movement in the Uttar Pradesh hills and the Silent Valley movement in the South were expressions of growing awareness about the environment. Later, two developments changed course of the history of forest management in India—one was the Forest (Conservation) Act, 1980 and the second was the Supreme Court's orders in the Godavarman vs. Union of India case. With the slowdown and even reduction of indiscriminate logging operations in some states and the beginning of the conversion of regulated forest

Table 1.1 *Forest Management in India, 1850–present*

1.	1850 to 1947	—	colonial exploitation
2.	1947 to 1985	—	continuation of colonial legacy
3.	1975 to 1990	—	conservation as priority
4.	1991 to 2000	—	political indifference
5.	1995 to 2010	—	judicial activism
6.	2010 to present (on going as of 2018)	—	(a) economic development overrides conservation ethos; (b) dilution of environmental supremacy

lands to non-forest uses, the business as usual might have paused, but the damage had already been done and forest degradation continued. Forests became degraded, conflicts over this resource were exacerbated and sociopolitical forces superseded conservation ideals, rendering sustainable forest management a delusion.

The preceding three decades (1950–1980) also witnessed a worldwide environmental movement, which sent reverberations into India as well. India had two prominent movements in this time that affected the way the environment and forests were looked upon.

The period also witnessed thousands of hectares of clear-cut forest areas bearing failed plantations of economically valuable and fast-growing species, for example, teak and eucalypts. The clear-cutting rendered biodiversity-rich forests into barren lands, suffering from heavy grazing and bearing only unpalatable scattered shrubs and grass tufts. One example of such an unsuccessful transformation was in the Panchmahals district of Gujarat. However, all states had this disappointing experience. A centrally sponsored scheme—Plantations of Economic and Fast Growing Species Schemes—provided funds to the states, which prompted the latter to clear-cut mixed-species, biodiversity-rich forests, removing all flora and planting that area with fast-growing exotic species. The proponents and implementing agencies of this scheme and similar other development schemes totally ignored the social and ecological considerations. When clearance of natural forests deprived local communities of access to nearby forests for grazing domestic livestock and extraction of fuelwood and small timber, they were forced to cut newly planted saplings year after year and the domestic cattle were let loose as their grazing grounds were lost. The forest establishment tried to enforce laws to prevent people from destroying plantations, but mostly failed. There was hardly any scientific basis for such a programme and approach, and nor did the foresters try to understand biological and ecological principles. They did not pay any attention to the outcome of such an activity and its impact on soil moisture and nutrients. Not that failure was everywhere; in many localities, successful plantations were raised. But there was failure too on a significant scale. Stocktaking was hardly done later and lessons were rarely learnt.

When the Indian Forest Service (IFS) was constituted in 1966 as an all-India service common to both the central and the state governments, and the first recruits came in 1968, the service was constituted by appointing eligible members of the state forest services, who were called 'initial recruits' and got benefit of seniority over the fresh new recruits hired through the Union Public Service Commission (UPSC) through open competition. However, even with the higher status of the IFS, no material change took place in the thinking and approach of the state forest departments and it was still business as usual. The direct recruits to the IFS were oriented and mentored by the initial recruits and promoted officers, and the former also (knowingly or unknowingly) got sucked into the system of subservience, meekness and a comfortable lifestyle. However, a few significant developments are briefly discussed below.

Silent Valley Movement

Save Silent Valley was a social movement that was aimed at the protection of the pristine tropical moist forests of Silent Valley in the Palakkad district of Kerala. A hydroelectric project was planned on the river Kunthipuzha that originates in the dense forests of Silent Valley. In 1970, the Kerala State Electricity Board proposed a hydroelectric dam on this river. According to the project report, it would submerge 830 hectares of primary forests. The project was approved in 1973 by the central Planning Commission. This led to a most active environmental debate and an agitation of a magnitude hitherto not seen. A non-governmental organization (NGO) named Kerala Sasthra Sahithya Parishad (KSSP) was the leading organization that was able to effectively mobilize other groups and the masses through a relentless awareness campaign about the value of biodiversity and conservation. Sugathakumari, a poet, also played an important role and even wrote poems on the subject. Scientific research by international and national scientists preceded this movement, and it was clear that the project threatened many rare and endangered species, such as the lion-tailed macaque, in addition to a number of endemic plant and animal species. Several prominent personalities of that time, including Salim Ali and M. S. Swaminathan, also wrote to the government to scrap the project. A writ petition was filed in the Kerala High Court, which was later dismissed. A committee under Professor M. G. K. Menon recommended the declaration of the Silent Valley as a national park. A long battle lasting more than a decade finally was won, the project was scrapped in 1983 and the area was notified as a national park in 1985. Conservation India carried out a detailed case study of this project that gives an account of the successful environmental battle.

Chipko Movement

In the mid-1970s, another conservation movement was simmering in the hill areas of erstwhile Uttar Pradesh (UP) which now make up the Uttarakhand state. The people living near the forests hugged or embraced the trees and did not let contractors from the forest department cut them down. The UP hills had been subjected to indiscriminate heavy logging, causing serious denudation and soil erosion. This finally resulted in a ban on commercial felling of trees growing above an altitude of 3,000 feet in this state. This movement was popularly called the Chipko movement, meaning 'embrace tree to resist cutting'. The resulting policy was later applied to Jammu and Kashmir and Himachal Pradesh as well, and under the Supreme Court's orders, there too the cutting of green trees was banned. These and other movements reflected resentment against the reckless policies that led to forest and biodiversity destruction.

Stockholm Conference

In 1973, the United Nations (UN) Conference on the Human Environment in Stockholm took place, where the international community agreed to protect the environment from pollution and forest degradation. In 1972, India had promulgated the Wildlife Protection Act and adopted the International Union for Conservation of Nature (IUCN) definition of protected areas. Very soon, wildlife sanctuaries and national parks were notified in the states. This would later on save certain areas from commercial exploitation and promote the preservation of ecosystems.

Forest (Conservation) Act, 1980

A very significant development took place in 1980, when the Forest (Conservation) Act (FCA) was promulgated on 25 October. It provided that no reserved forest shall be de-reserved or used for non-forest purposes without the prior approval of the central government. This policy approach had been first communicated to the state governments through guidelines which were however never implemented. The states enjoyed total freedom to clear and distribute forest land for agriculture or any other use. The FCA, in effect, took away that power from the states. The FCA aimed at preventing indiscriminate clearance of forests or their conversion to other land uses. The FCA has been the most reviled central law till date, though it has helped save millions of hectares of reserved forests from deforestation.

Recently, the aforesaid law has been diluted. To expedite the grant of forest clearance to linear projects such as roads, railways, canals, transmission lines and pipelines, most of which are of a public utility nature, the Ministry of Environment, Forest and Climate Change (MOEFCC) has decided to delegate the powers to grant forest clearance to such projects, irrespective of the area of forest land involved, to the Regional Empowered Committee being constituted at each regional office of the MOEFCC. The MOEFCC has also issued guidelines that in case of linear projects, 'in-principle' approval under the FCA may be deemed as the working permission for tree cutting and commencement of work, if the required funds for compensatory afforestation, net present value (NPV), wildlife conservation plan, plantation of dwarf species of medicinal plants and all such other compensatory levies prescribed in the in-principle approval are realized by the user agency (MOEFCC guidelines, 2 September 2014).

Social forestry and afforestation

In the mid-1970s, another interesting movement of tree planting was initiated that gave a boost to forestry establishment. This later on led to social forestry and afforestation on a mass scale—beginning in the late 1970s, afforestation was significantly scaled up in the decade of the 1980s. Afforestation was included in the 20 Point development agenda in 1981 and was monitored on a regular basis by the central government. This was possible only with political support. The budget for afforestation was also included in two centrally supported schemes, the National Rural Employment Programme (NREP) and the Rural Landless Employment Guarantee Programme (RLEGP), thereby making it obligatory to allocate 25 per cent of the funds for afforestation activities. International donor agencies—both multilateral, such as the World Bank (WB) and the European Union (EU), and bilateral, such as the United States Agency for International Development (USAID), the United Kingdom Overseas Development Authority (now UK Aid), the Swedish International Development Cooperation Agency (SIDA), the Canadian International Development Agency (CIDA) and the Overseas Economic Cooperation Fund (OECF) of Japan (now the Japanese International Cooperation Agency, or JICA)—jumped on to this bandwagon, bringing millions of dollars.

Ambitious targets for afforestation were set by the central government for each state, irrespective of its budget or the capacity of its forest agency. At the same time, a huge amount of funds from both domestic and external entities started flowing. Social forestry departments were set up in most states. This programme also raised awareness among local people and encouraged their participation, though through employment alone to begin with. The benefits that flowed to the local communities were in the forms of wage employment and increased supply of grass fodder for domestic cattle.

The criticism of the programme was that it encouraged malpractices. There were allegations that inflated progress was reported, all saplings planted did not survive, agroforestry achievements were exaggerated and the results did not correspond with the investments made during the whole decade. A large number of foresters however contended that it was social forestry that saved the natural forests by providing wood from the trees growing outside the legally designated forests. Some critics of the agroforestry programme still argue that it was not the forest departments that promoted agroforestry, but rather it was the demand for industrial woods such as poplars, eucalypts or acacias that prompted the private landowners to plant trees on their farmland in blocks or strips along farm boundaries.

However, the fact remains that the recorded figures on afforestation, social forestry and agroforestry did not match those really established on the ground. The success was far less than claimed. According to a government report, between 1979 and 1997, an area of 25 million hectares was planted with trees (GOI 1999). This could not be supported by the ground realities, which indicates that it was an inflated reporting. The reports were never corroborated by comprehensive monitoring and evaluation. While social forestry was becoming popular as the flagship policy and programme of India's forestry sector, the degradation of state-owned forests was progressing as usual at an alarming rate. The extraction of wood was not followed by the availability of financial resources for reforestation.

CHAPTER 2

Forest Policy

A policy for a specific sector of the economy is aimed at the development of that sector, at improvement in productivity and growth. If we take any policy of a sector—an industrial policy, an agricultural policy or an educational policy, for that matter—we would find that the policy has supported growth and development in that sector. For example, the agricultural and industrial production have increased over the years, even if targets were not achieved fully, and literacy has improved, more educational institutions have come up, and so on. It is only the forest sector that has not only stagnated but has shown negative growth. It has not been able to achieve the policy goals set in either 1952 or 1988.

Colonial Forest Policy

The Forest Policy of 1894, logically, stipulated that the claim of agriculture should prevail over forest-land use. It stated:

> It should also be remembered that, subject to certain conditions to be referred to presently, the claims of cultivation are stronger than the claims of forest preservation. The pressure of the population upon the soil is one of the greatest difficulties that India has to face, and that application of the soil must generally be preferred which will support the largest numbers in proportion to the area.

Accordingly, wherever an effective demand for cultivable land exists and can only be supplied from forest areas, the land should ordinarily be relinquished without hesitation, and if this principle applies to the valuable class of forests under consideration, it applies a fortiori to the less valuable classes which are presently to be discussed. When cultivation has been established, it will generally be advisable to disforest the newly-settled area. But it should be distinctly understood that there is nothing in the Forest Act, or in any rules or orders now in force, which limits the discretion of Local Governments, without previous reference to the Government of India (though, of course, always subject to the control of that Government) in diverting forest land to agricultural purposes even though that land may have been declared reserved forest under the Act.

National Forest Policy, 1952

Though this policy enunciated that forest land should not be cleared for agriculture, between 1950 and 1980, 2.6 million hectares of forests were converted to farmland besides the 1.7 million hectares cleared for mining, infrastructure, industry and urbanization. Thus, officially, 4.3 million hectares of forests were lost on authorization by the government and more than 2 million hectares were cleared illegally, converted for agriculture and other activities during the three decades since the inception of this policy. This is one piece of evidence of policy failure. In 1956, the Grow More Food campaign also encouraged forest conversion. States distributed title deeds over forest lands after encroachment, a big political tool for votes. Also, overexploitation and consequent depletion of forest resources continued at an accelerated speed. The states were not seriously committed to the implementation of the forest policy and the policy was merely a piece of paper, an aspirational statement that could be quoted in essays and in academic and technical literature.

To a large extent, the 1952 policy was a continuation of the colonial policy, except that it envisaged that forest land should not be converted to agricultural land, which was encouraged in the previous policy of 1894. Too much emphasis was laid on industrial and national needs, implying increased harvesting and increasing yield of timber. It did not take into consideration the dependence of local communities on forests in their vicinity. The 1952 policy also put down sustained yield management as a principle of forest management. Yet vast tracts of forests were exploited indiscriminately for timber and fuelwood harvests. No systematic monitoring was done of forest resources, but as revenue from the forests was a priority and conservation a principle, revenue considerations superseded the conservation ethos. The policy reflected the opinions and aspirations of the actors in the forest sector and it appeared to have not taken on board the views of other stakeholders. This resulted in a narrow vision and an unrealistic approach, which were bound to lead to disastrous outcomes.

The policy could not achieve its objectives. The condition of the forests deteriorated during the operational period of this policy. One third of the country's land area could not be brought under forest cover as stipulated in the policy. On the contrary, 4.3 million hectares of forests were cleared during the first 30 years of the policy for agriculture, mining, urbanization, irrigation and other infrastructure development. Sustained yield forest management did not become successful.

Forest (Conservation) Act, 1980 (FCA)

In 1980, the FCA was promulgated to restrict indiscriminate use of forest land for non-forest uses. This was a break with the 1976 constitutional amendment that brought forests under the concurrent list. The FCA had been preceded by guidelines in 1978, placing restrictions on deforestation. These guidelines were neither respected nor implemented. The FCA could not eliminate deforestation; only, the power was transferred to the Government of India (GOI) and the procedures made very cumbersome to discourage their indiscriminate use. However, the FCA did introduce regulation of forest-land diversion, though it merely took away the power of diversion of forest land from the states. The FCA has been the most despised piece of legislation by all state governments, bureaucrats and politicians. Its implementation on the ground has involved serious conflicts between the forest department and all the other departments. Yet, there has been serious deforestation and degradation in the period of its application.

National Forest Policy, 1988

The objectives of India's National Forest Policy (1988), which supposedly adopted a new strategy for forest conservation, can be grouped under the following five themes:

(a) Conservation of forest resources, the environment, wildlife and soil, and prevention of desertification
(b) Improvement of forest productivity
(c) Development of resources by afforestation and reforestation, improvement of productivity and efficient utilization of forest produce and promotion of wood substitution
(d) Meeting the demands of local communities for fuelwood, fodder and small timber and of the essential needs of the nation

The 1988 policy stipulated that to achieve its objectives, peoples' cooperation should be enlisted for the protection and management of forests by creating a mass movement.

Conservation, as interpreted in the National Forest Policy of 1988, included preservation, maintenance, sustainable utilization, restoration and enhancement of the natural environment. Thus, the conservation of the forests remained the dominant spirit of the policy. The policy did not prohibit sustainable use of the forest resources. It did call for preservation of forest areas rich in biodiversity. It sought to ensure environmental stability. The protected areas required a greater degree of protection. The policy articulated a principle that economic considerations would be subservient to environmental considerations.

The objective, though with progressive dilution, of restricting diversion of forest land for non-forest purposes had been achieved to some extent through the FCA of 1980, which in fact came before the policy of 1988. The 1988 policy simply endorsed the FCA, which was amended in 1988 along with the promulgation of the new policy to give it more teeth.

The raw material supply to wood-based industries at concessional rates was phased out, which was an important policy stipulation.

Afforestation and social forestry were supported, which later declined in the 1990s. Biodiversity conservation also got an impetus. Many of the initiatives included in the policy were already under implementation when the policy was adopted, though.

However, there is no evidence on record to establish that the objectives of conservation of natural heritage (fauna and flora), prevention of soil erosion, prevention of denudation of catchment areas, prevention of desertification and conservation of water have been achieved. The protected area (PA) management has come under controversy from time to time. The state forest departments (SFDs) did not complete the process of notification of PAs under the Wildlife (Protection) Act, 1972, and the rights of local communities were not settled. As a result, the final notifications could not be issued. There was a rush to notify more and more areas as PAs under the Act. The basic criterion adopted for declaration of PAs was the presence of big mammals such as tigers, lions, elephants, rhino, bison, blackbuck and chinkara, though some PAs had birds, such as Keoladeo Ghana. Not all PAs demarcated on the basis of ecological significance took into account an ecosystem as a whole or a landscape; rather, many took into consideration only fauna, and that too, big mammals, earmarking mainly areas used as hunting grounds during the colonial era. Also, the management of PAs so notified has not been effective. The more popular programmes such as tiger and elephant conservation were given priority. Forest conservationists argued that by protecting the tiger at the apex of the food chain, the whole ecosystem would be conserved. Over the years, though, the number of tigers have decreased, and so have lions and other animals.

Despite huge expenditure on the preservation of tigers under the Project Tiger programme, their population has declined during the last 40 years since the flagship programme was taken up. Similarly, the Asiatic lion in the Gir forest has also showed a decline in population and even stands in danger of extinction. The PAs have been set up and are being managed as islands. The government did not take adequate measures to create networks, corridors and connectivity. Scientifically, the biodiversity conservation objectives would have best been pursued if a landscape approach had been adopted. There is a serious issue of governance at play here. Policies are written and actions are initiated halfheartedly. A programme is started but not sustained and is not brought to completion due to either lack of commitment or resources, or both. Historically and popularly, by 'wildlife' everyone meant only wild animals and, therefore, a holistic approach to biodiversity conservation on an ecosystem basis has been rather rare.

Political and economic compulsions and expediencies, litigations and NGO activism have discouraged effective implementation of conservation policies, including the National Wildlife Action Plan. Conservation efforts have also involved serious conflicts with local communities and other stakeholders, as well as within forestry establishments. For example, in Jammu and Kashmir, Dachigam National Park, which is a habitat of the rare, endangered and endemic Kashmir stag (*hangul*), allows for grazing and seasonal settlements of shepherds with about eighty herds of sheep and goat and also has a government sheep-breeding farm. This is detrimental to *hangul* breeding and the survival of its fawns, which become easy prey to the sheepdogs kept by the shepherds. The *hanguls'* grazing areas are restricted and so is their free movement. The politicians prefer sheep to *hangul* and in this conflict, the latter is losing. The national park managers are silent spectators as any actions to set things in order will result in their transfer to other posts.

There is a serious man–animal conflict. The profusely growing, unsustainable human population is intruding in and gradually destroying wildlife habitats. The conservation lobby has become silent and weak. Biodiversity sacrifices for the sake of infrastructure development and economic growth are the norm these days in India.

Expansion of Forests

The goal of achieving 33 per cent of the country's land being under forest cover continues to remain a distant dream. The policy does not give a reason for having one third of country's land area under forest cover. Nor does it provide guidance as to where and how this expansion would take place. The rhetoric or aspiration, being reiterated in the policy circles for the last seven decades, is so unrealistic that it is ridiculed in international circles and it would have been advisable to abandon it a long time ago, or it should not have been included in this policy at the first instance. If one looks at the periodic forest-cover monitoring reports of the Forest Survey of India (FSI) since 1988, it can easily be observed that there has been no significant increase in the forest cover in India to move towards the goal of 33 per cent forest cover. Also, looking at the trend where the existing forests have not been properly maintained and managed on a sustainable basis, how could the principle of forest expansion be justified or be an overarching goal? It is unrealistic.

Private industry and forest raw materials

Prior to 1988, a large number of wood-based industrial units were dependent upon the state governments for supply of raw materials at a throwaway price. Many of these concessions were political favours and patronage. The states had entered into long-term contractual arrangements for the supply of wood and bamboo in some states. The 1988 policy stipulated that the practice of supplying forest produce as raw material to industries at concessional prices should cease. The industry should procure raw materials from farmers or from imports instead. Some Himalayan states had long-term agreements for the supply of resin, some for timber for joinery and the plywood industry and some for bamboo to rayon and paper mills. Following the 1988 policy, most states phased out the concessional supply of raw materials and no new contracts were entered into, either at their own instance or under pressure from donor agencies or the courts. The farmer–industry relationship has flourished over the years. Farmers are growing trees and supplying raw materials for pulp, paper and plywood industries and even for furniture and house construction activities. Huge quantities of sawn timber and furniture are being imported from Southeast Asian and other countries. The negative impact has been that there is no compulsion on the part of the government to improve the productivity of forests to meet local demand. It is also a fact that locally produced timber from state forests is more expensive than that imported.

The lobby for the wood-based industries has been pushing for a long time for the leasing out of forest land for production of timber and other forest products to provide captive raw materials. It has argued from time to time that if land is purchased and capital is borrowed from a financial institution, an industrial plantation project would financially be unviable. The politicians have also been influenced from to time to buy this argument and go for such a scheme. Even recently, there has been discussion on a policy to facilitate private entrepreneurs by leasing forest land at a nominal rent for reforestation and afforestation. In 1986, the National Wasteland Development Board (NWDB) mooted such a policy and came up with a proposal. This was met with strong opposition from civil society and from within the government. This policy has never been introduced yet but there is suspicion that it may be initiated again if vested interests get an upper hand.

If such a policy is adopted and implemented, it will be disastrous for the country. It will involve privatization of common property resources that are meant for the public good and the industry will only grab land that will gradually undergo land-use changes due to loopholes and non-enforcement of safeguards. Once ownership has been transferred even on a long-term lease basis, the enforcing public agencies will withdraw. Moreover, the industry wants productive land and not forest land that is degraded and eroded. Some private companies that promoted tree plantation schemes for timber production, showing highly attractive returns on investment, duped thousands of people of their investments and went bust, with their owners chased and prosecuted for cheating. Forest plantations in India cannot be a profitable business if land is bought or taken on rent. Farmers raise plantations on their land in agroforestry models on farm bunds and vacant land without jeopardizing their agricultural production, or absentee landlords put their land under tree plantations to get a passive income. These lands have access to irrigation and fertilizers and the tree growth is good, ensuring attractive returns on short-rotation tree species such as eucalypts, acacias and poplars. This author conducted a financial analysis and found that when rent of land is added to the cost and a commercial rate of interest is used for discounting, a tree plantation project would have a negative net present value (NPV).

Monitoring of Policy Implementation

The policy implementation has not been regularly monitored. It was, however, reviewed by a committee in 1999, which recommended not to change it. Since no baseline survey was done to monitor the policy implementation, it would be impossible to make any empirical assessment of the changes that could be attributed to the policy. The fact remains that degradation of forests continued and so also decrease in forest biodiversity, in and outside the PAs. The dependence of rural people on fuelwood, fodder, small timber and grazing continued, which has also been driving degradation. Both illegal occupation of forest land and official conversion of forests to non-forest uses, such as mining and infrastructure development, continue. There exists no tenable evidence that the policy has led to the enhancement of conservation.

Outcomes

A detailed analysis of the policy implementation, outcomes and constraints are given in a matrix in the Appendix. The 1988 policy could not be effectively implemented because of being unrealistic and not taking into account ground realities and the lack of political commitment and institutional constraints. The conflict on land tenure and use could not be adequately addressed and a wish was expressed regarding the regulation of grazing by limiting it to carrying capacity. Saxena (1999, 44) believed that there were contradictions within the National Forest Policy of 1988 and between the policy and the FCA as amended in 1988, relating to the prohibition of horticultural crops, and medicinal and oil-bearing plants; leasing of land to people; granting of rights and concessions; exploitation of non-timber forest products; and tree felling on private land. He further contended that the IFS had reservations about the 1988 policy, though it had been more in alignment, in principle at least, with the previous two policies.

The GOI has admitted that there have been failures to achieve some policy objectives due to ineffective implementation. It has argued that it was not easy to follow the precepts of the forest policy due to widespread poverty in rural areas and that the policy basically indicated intentions only. Neither does the policy set quantitative or qualitative targets for achieving its broad objectives, nor does it say how long it would take to bring one third of the country's land under forest cover (GOI 2007b, 133). This means that the policy has only remained on paper and was not a prudent and implementable instrument, and therefore failure was inevitable.

If empirical assessment was impossible, at least a qualitative assessment could be carried out to see the outcomes of the three decades old policy. There is no denying the fact that many policies are general statements or a wish list that includes and reflects the desires of the stakeholders. If policy goals and strategies to achieve the set goals are unrealistic, it would appear futile to base sectoral behaviour on the policy precincts. Yet the GOI has never made an attempt to develop a set of even qualitative indicators against which the policy performance could be evaluated.

Opportunities for Policy Review and Change

National Forestry Action Programme

The National Forestry Action Programme (NFAP) was developed with the assistance of the United Nations Development Programme (UNDP) and the UN Food and Agriculture Organization (FAO). The work started in 1994 and was completed in three years. Under guidance of the central government, the states also developed state forestry action plans for their respective states. The NFAP was approved in 1999 and was developed for the next 20 years. The raison d'être for NFAP was that this comprehensive strategic plan was prepared by the GOI to address the issues underlying the major problems of the forestry sector, with the objectives of reversing the process of forest degradation and of promoting sustainable development of forests (GOI 1999).

The NFAP claimed that forests contributed 1.7 per cent to the gross domestic product (GDP) and were under tremendous pressures from large human and cattle populations. These pressures had resulted in deforestation and forest degradation. The main causes for this forest trend were shifting cultivation, unregulated cutting for fuelwood collection, livestock grazing, forest fires and diversion of forest land. The NFAP stated that the rate of officially authorized deforestation reduced considerably after 1980, as shown in Table 2.1.

The NFAP also stated that between 1950 and 1997, forest plantations were raised over 28.23 million hectares and the periodwise achievement was as shown in Table 2.2.

The NFAP acknowledged that the performance of the forest plantations in India, in terms of survival and yield, had been poor. Due to this, the effective area of forest plantations was estimated to be about 11 million hectares, that is, less than half of the reported area. It also admitted that the condition of several PAs was poor, though there were 80 national parks and 441 wildlife sanctuaries covering 14.8 million hectares of forest land.

There had been huge gap between demand and supply of fuelwood and timber per 1996 figures. Against a demand of 201 million metric

Table 2.1 *Deforestation, 1970–1997*

Period	Deforestation
1970s	1.3 million ha
1980s	339,000 ha
1991–1997	129,000 ha

Table 2.2 *Tree Plantation, 1970–1997*

Period	Plantations raised (million hectares)
1950–1980	3.54
1980–1990	13.51
1990–1997	11.30
————	————
TOTAL	28.35

tonnes of fuelwood, only 115 million tonnes was available from state forests and non-forest lands, thus leaving a deficit of 88 million tonnes. Similarly, against 64 million cubic metres of timber demand, the state forests were yielding (1996) only 12 million cubic metres and the non-forest land 31 million cubic metres, thus leaving a gap of 21 million cubic metres. This gap further increased after the implementation of the Supreme Court orders restricting commercial harvesting in state-owned forests.

The NFAP estimated a requirement of US$32 billion (₹1,340 billion) at the 1999 price level. At 2018 level, it could easily be multiplied by five after two decades to make it ₹6,695 billion (equivalent of US$106 billion). Five major components included in the NFAP were forest protection, improvement of forest productivity, expansion of forest area, reduction of total demand for forest products and strengthening of forest policy and the institutional framework. It was proposed in the NFAP that the finance for implementation would come from domestic sources as well as external sources. Other sectors such as rural development and agriculture would also provide funds, as also the private sector. The authors suggested that the actual mechanism for finance mobilization had to be designed and developed within the country (MOEF 1999b). Based on the NFAP, the Indian Ministry for Environment and Forests (MOEF, now MOEFCC) prepared a number of 3–4 page project profiles, put these in ring binders and distributed them among donor agencies' offices located in New Delhi. There was no further follow-up. These documents, after some time perhaps, were shredded by the donor agencies and the remainder gathered dust in the shelves of the MOEF office, to be weeded out at a future date when all have forgotten about this.

It was a strange governance performance. The state forestry action plans never saw the light of the day and the NFAP proved to be still-born. It was undertaken in 1994 as a highly euphoric, ambitious and fantastic activity by the MOEF officials but ultimately the job was completed with the printing and distribution of the document. This is evidence of how far removed the policymakers and planners were from the realities. The cost that was incurred both by the MOEF and the SFDs in terms of time and salary was substantial, besides about

US$700,000 of UNDP and FAO contributions and efforts. The NFAP goals and targets were totally unrealistic and infeasible. The states looked to the central government for next steps and finance, but little did they know that the MOEF's job was over—the officials who had steered the process got transferred and their successors lacked the institutional memory and hardly found time or interest to even revisit the well-written NFAP document, printed on glossy paper, waiting to be acted upon by someone who seriously and sincerely took follow-up action. It is not uncommon and many such efforts have met their fate in this manner.

There are a number of similar examples which prove that governance has been the main issue in the forestry sector and that its actors never succeeded in recognizing the realities to develop the right kind of strategies and act upon them. Poor governance brings frustrations and even those who want to do something feel demoralized and marginalized. The Indian bureaucracy, which inherited the colonial mindset and added to it a feudal temperament, could have delivered much more for the country if it could have been sincere, committed and forthright and if it had professional integrity. But its members always focused on their self-interest in terms of postings, for which they found no alternative but to align with and serve the interests of self-serving power centres. The forest officers are also part of this system. The 'system' is so powerful, overwhelming and deeply entrenched that it acts ruthlessly against non-conformist elements and constructive dissent.

National Forestry Commission

The Indian Board of Wildlife (IBWL), headed by the Prime Minister of India, in its meeting on 21 January 2002, resolved to constitute a National Forestry Commission (NFC) to review the working of the forestry and wildlife sector. The IBWL resolution emphasized the paradigm shift in the tenets of forest management from the primacy of timber to ecological and stakeholder-oriented forestry, taking cognizance of the recommendations of the National Forest Policy (1988), the Stockholm Conference (1972), the United Nations Conference on Environment and Development (UNCED 1992) and the continued pressure on forests despite the enactment of the Wildlife (Protection) Act (1972) and the FCA (1980).

The MOEF took a year to constitute the National Forestry Commission (NFC) by a notification that was finally issued on 27 February 2003. The main terms of reference for the NFC were:

(a) Review and assess the existing policy and legal frameworks and their impact in a holistic manner from ecological, economic, social and cultural viewpoints
(b) Examine forest administration
(c) Suggest policy options for sustainable forest and wildlife management
(d) Suggest to make forest administration more effective
(e) Establish relationship between forest management and communities.

The NFC, headed by a retired Chief Justice of the Supreme Court of India, B. N. Kirpal, included the Director General of Forests (DGF), NGO activists, academics and the Additional Director General of Forests as its member secretary. The NFC submitted its report in March 2006, after working for about three years. It is not useful to go into the details and methodology of its working. What we should focus on is its main findings and recommendations. The report of the NFC is a voluminous document, running into 421 printed A4-size glossy pages and an attractive cover. Four of its members—J. C. Kala (the DGF), G. K. Prasad (Additional Director General of Forests), A. P. Muthuswamy and Chandi Prasad Bhatt (a leader of the Chipko movement)—submitted dissent notes, which were included in the annex to the report.

The NFC made 360 recommendations and concluded that there was no need to amend the 1988 Forest Policy and the objective of bringing one third of the country's land under forest cover. It recommended that the Indian Forest Act of 1927 be amended and that the FCA of 1980 should not be diluted. The NFC emphasized that the main objective of forest management should be ecological security. Its recommendations also included addressing the livelihood needs of forest-dependent communities, sustainable management of forests, regulation of grazing, afforestation of 29 million hectares of land, fire protection measures, control of exotic species and alien invasive species, and more (GOI 2006).

In the ultimate analysis, the recommendations of the NFC were highly generalized and non-controversial. When the report was made public, it did not invoke any perceptible excitement among stakeholders. The report yet formed a decent document. Its greatest weakness, however, was its excessive size and length and colossal number of recommendations. Certainly, it was an exhaustive document. However, a focused and crisp report could have been very useful for the forestry establishment, policymakers and other stakeholders. One issue that missed the attention of the NFC was that of land tenure (forest villages, rights of tribal communities and other forest dwellers), which was not examined and discussed in the report. In the same year, 2006, the central government promulgated the Scheduled Tribes and other Traditional Forest Dwellers (Recognition of Forest Rights) Act, 2006—called the Forest Rights Act (RFA) in popular parlance—to regularize the rights of the forest dwellers on forest land under their possession.

New policy under preparation

If the central government is not inclined to accept failures in implementation but has therefore decided to develop a new policy ('revision' would be an inappropriate word to use for a 30-year-old document), it indicates that the existing NFP has lost its relevance in the present socio-economic and global context. It also reflects that the institutions which prepare and implement policy documents do not have adequate capacity and capability to carry out policy evaluation, analysis, preparation and development of strategies and action plans for policy implementation and monitoring. An absence of concomitant strategies and action plans to achieve policy objectives with well-defined input, output and outcome indicators has been a major reason for the lack of feedback and failure of the learning process.

When the preamble to the new forest policy of India that follows the 1988 policy is finally written, one wonders what it will be. In all likelihood, the bureaucrats will interpret the previous policy to have been successful and list positive outcomes, and may show there was forest cover increase, Joint Forest Management (JFM), and so on. Many of the positive outcomes are a result of the Supreme Court's directives from time to time and external donor-driven changes in terms of the phasing out of subsidy, easing harvesting and transit regulations, new technology and capacity building.

The Indian Institute of Forest Management (IIFM), Bhopal, prepared a Draft National Forest Policy, 2016, which was placed on the website of the MOEFCC. Later on, it was termed as a report of a study of the existing forest policy. The MOEFCC has since posted on its website (www.envfor.nic.in) a Draft National Forest Policy, 2018, that is claimed to have been prepared by the MOEFCC after stakeholder consultations.

The 2018 draft is a vague hotchpotch of unrealistic and ambiguous ideas, with confused goals and objectives, yet full of rhetoric. It shows a total disconnect from socio-economic and ground realities. The overall goal of the draft policy is to safeguard the ecological and livelihood security of the people, of the present and future generations, based on sustainable management of the forests for the flow of ecosystem services. For achieving this goal, the country should have at least one third of its land area under forest cover, according to the policy. The draft is similar to the existing policy documents except the addition of jargon such as 'climate change', 'REDD+', 'sustainable', 'participation', 'technology', and so on. The authors have no clarity on concepts and approaches that should be adopted for the development of a realistic policy framework.

The traditional opinions, constricted thinking and indifference to an analytical approach present insurmountable constraints. Given the authors' rigidity and egoistic overconfidence, with little scope of assimilating ideas from various other members of civil society, the possibility of a sound draft is low. The ritualistic stakeholder consultations would not be useful beyond complying with formalities. I predict that when the draft undergoes churning from other centres of power that have a greater say within the government system, it may not receive a positive or near-positive response and far-reaching changes will be made, rendering the draft an ambiguous or generalized, redundant statement, unqualified to satisfy various actors from different sectors and levels.

There should be a third-party policy analysis and formulation agency having experts from various disciplines such as economics, sociology, environment, wildlife, forestry, climate-change science, hydrology, energy, agriculture, mining, irrigation, finance, management, human resource management, law, communications and media. This can obviate the limitations involved in a purely bureaucratic discourse on policy issues and thereby prevent the emergence of a narrowly conceived policy document.

CHAPTER 3

External Donors Influencing Policy
The World Bank

Externally Funded Social Forestry Programmes

External donor agencies have provided about US$400 million for implementing social forestry programmes in 15 states of India. The bulk of the funds came from the World Bank followed by the Swedish International Development Agency (SIDA). The social forestry contributed to increased biomass production, farm forestry expansion and awareness for growing trees. Its negative contributions included large-scale corruption and misappropriation of project funds by fake plantations. It also resulted in neglect of core forestry sector, that is, management of natural forests. It did not help in introducing significant reforms and institutional development in forestry sector.

Donor Involvement in Social Forestry

The donors that provided substantial financial assistance for forestry included United States Agency for International Development, the World Bank (WB), the Swedish International Development Cooperation Agency (SIDA), the Canadian International Development Agency

(CIDA), the United Kingdom Overseas Development Administration (ODA), the Japan Bank for International Cooperation (JBIC) and the European Union (EU). Their involvement can be divided into three contiguous, and at the same time distinct, phases:

(a) Phase I (1979–1991): Social forestry programmes—supported by the World Bank, the United State Agency for International Development (USAID), UK ODA, SIDA and CIDA
(b) Phase II (1991–2002): Forest-sector reform and participatory forestry—supported by the WB; afforestation and plantation—supported by Japanese aid (the Overseas Economic Cooperation Fund, or OECF, and the JBIC)
(c) Phase III (2002–2018): Forest-based livelihood programmes—supported by WB, JBIC/JICA; 2008 onwards, mostly JBIC and small grants from the Global Environment Facility (GEF)

With social forestry at its peak, attention came to the core forestry sector—mainly a reorientation driven by WB experts, who saw social forestry depriving the core forestry sector from having financial resources to invest in the maintenance and restoration of the productivity of natural forests. Social forestry was also coming into disrepute due to lack of accountability, with the public doubting that all was not well, that actual afforestation was much less than reported and that fraudulent reporting could not be ruled out. Use of poor quality of seedlings and poor plantations raised were commonly seen. No attention was paid to improved plantation techniques, site–species matching or adequate planning, and the whole activity was target-driven. The focus was on quantity and not on quality.

In the mid-1980s, there were conversations on building a people's movement for afforestation. However, at the ground level, there were no community institutions that would participate in forestry activities. In many states, village-level forest committees were set up and in some villages, the panchayats (village councils) were involved in the plantation on the common lands, panchayat land and other community lands.

The WB changed its forest policy in 1991, which led to the beginning of a new generation of forestry projects in India. The Maharashtra and West Bengal forestry projects which were launched in 1992 still carried the legacy of social forestry and included components which were implemented through the earlier social forestry projects. The West Bengal project aimed at expanding joint forest management, piloted by the forest department in Arabari village.

Later, this participation model was adopted in the regeneration of degraded forests through a circular issued by the central Ministry of Environment and Forests (MOEF) in 1990. This arrangement was later called 'joint forest management (JFM)' under which the local participating communities were assured a share in the forest produce, either 25 or 50 per cent. JFM originated in a small village Arabari in West Bengal in 1972, as an experiment that helped in the regeneration of sal forests with the cooperation of local villagers, by protecting the coppice growth of sal in their vicinity.

The World Bank

For three decades from 1979, the WB was the single largest donor agency providing financial assistance for forestry programmes in India. The International Bank for Reconstruction and Development (IBRD), later called the World Bank, was set up in 1945 with the objective of assisting the reconstruction of the European economy ravaged by the World War II. It later extended its operations worldwide, providing development assistance to less- and underdeveloped countries. In India, the WB's lending operations involved both IBRD loans and financing by the International Development Association (IDA), a soft credit window of the IBRD. The IDA credit had a repayment period of 35 years, with a 10-year grace period and a service charge of 0.75 per cent annually on the amount withdrawn, plus a 0.50 per cent commitment charge on the amount, though obligated but not withdrawn.

World Bank-Assisted State-Level, Sector-Wide Forestry Projects

With the launch of India's Forest Policy of 1988, which replaced the policy of 1952, a new direction was given to forestry and a new thrust was given to conservation and community involvement in forest management. The WB discussed with the Government of India (GOI) in 1989 how to shift the focus of WB assistance from interventions in the social forestry sub-sector to a more comprehensive sector-wide approach, with a view to providing much-needed investment in the core forestry sector, that is, in natural forests. Despite the emphasis on social forestry and increased wood supply, India's ever-increasing population and economic development continued to put greater pressure on forest resources. This led to the preparation of a new generation of forestry projects with WB assistance, starting with Maharashtra (Cr. 2528-IN, US$124 million) and West Bengal (Cr. 234-IN, US$34 million, closed in 1997) forestry projects, both launched in 1992. These were followed by an Andhra Pradesh forestry project (Cr. 2373-IN, US$77.4 million, in 1994), a Madhya Pradesh project (Cr. 2700-IN, US$58 million, 1995), an Uttar Pradesh project (US$58 million, 1996), a Kerala forestry project (Cr. 3053-IN, US$39 million, 1998) and an Andhra Pradesh Community Forest Management Project (credit of US$106 million, 2002).

The main objectives of the WB's forestry portfolio were:

(a) Community participation in regeneration, development and protection of forests through a joint forest management approach, involving sharing of usufruct
(b) Institutional and policy reforms and capacity building of public forestry institutions, village communities and NGOs
(c) Upgradation of technology for management of forests and plantations and for improving productivity
(d) Conservation of biodiversity in the designated protected areas through improved management, community participation, awareness raising and research and studies to have a better understanding of specific ecosystems and biomes.

Outcomes

(a) Conditionality used for reform. The conditions of financial assistance leveraged institutional reforms in the project state
(b) Policy reform under WB-assisted projects

(c) Support to implement the National Forest Policy of 1988
(d) Leverage to introduce changes in partnership with the forest departments
(e) Felling and transit regulations
(f) Long-term concessions
(g) Revolving fund set up in Andhra Pradesh and Kerala
(h) *Nistar* (rights to get free forest produce) in Madhya Pradesh restricted to 5 kilometres radius
(i) Development of state policies
(j) Biodiversity strategy
(k) PA network rationalization
(l) Information system strategy
(m) Subsidy on seedlings
(n) Institutional studies
(o) Restructuring of state forest departments in West Bengal, Andhra Pradesh, Maharashtra and Kerala.

The restructuring in West Bengal was a blunder, forced through legal covenants and the threat of suspension of credit (loan) disbursement. It was allegedly aimed at helping a certain cadre of employees.

The World Bank and Controversies

The WB provided assistance for irrigation and agricultural projects from the 1950s onwards and extended support for forestry programmes in the late 1970s, commencing with social forestry projects in Uttar Pradesh and Gujarat. In the 1980s, the Sardar Sarovar irrigation and hydroelectricity project in Gujarat became a subject of controversy and gave the WB jitters. This project was investigated by an inspection panel (Morse Commission) set up by the WB, which found that the project would adversely affect thousands of indigenous (tribal) people, displace them from their land, snatch away their livelihoods and violate human rights.

Some NGO activists in India, prominent among them being Baba Amte and Medha Patkar, led agitations against the GOI and the WB. The WB management, as they always do, withdrew from the project and cancelled the loan in consultation with the GOI. Thereafter, the WB developed very stringent policies and procedures, called the Environment and Social Safeguards, and organized a new and powerful division—the South Asia Social and Environmental Division (SASED)—which gradually assumed the role of policing to ensure compliance with the WB safeguard policies and would subsequently put on hold many of the WB's projects after preparation of reports and even after final appraisal for lending. The operational staff and even senior managers in the WB became terrified of this division, which reported directly to the regional vice president. A number of SASED staff in India were recruited from activist NGOs and were anti-establishment, with a contempt for the government and for government agencies, and had a tendency not to pass up any opportunity to embarrass the government and put unreasonable conditions upon project finance under the garb of safeguard policies.

World Bank Aversion to Forestry Institutions

After the India—Ecodevelopment Project was scrapped (see Box 3.1), the WB faced another controversy while funding the Madhya Pradesh Forestry Project. Some NGOs that called themselves mass and tribal organizations were at loggerheads with the government of Madhya Pradesh on the issue of forest-land distribution to tribal communities and giving title deeds over the forest land that had long been cultivated by the tribal villagers to those who were brought on for logging operations and required land for dwelling and cultivating, and also titles to the land they had encroached upon. There were serious conflicts and even violence when the forest department attempted to evict encroachers or prevent encroachments. The tribal advocates' argument was that the tribal communities had their primitive and traditional rights over all forests snatched away from them during colonial rule. On the other hand, the forest department argued that they were only enforcing the forest laws, for they were held accountable by the government. The principle which the tribal activists propagated was that water, land and forests belonged to the tribal peoples.

Some of these organizations had friends in the WB's New Delhi office and sought their advice on how to file a complaint with the WB

Box 3.1 The World Bank

The WB supported India's forestry sector for more than three decades. In the 1980s, it provided financial assistance for forestry projects in nine states. From 1991, the WB shifted its support to state-level, sector-wide, integrated forestry projects in six states, with an innovative repeater project in Andhra Pradesh. In addition, there was one project each for forestry research implemented by the Indian Council of Forestry Research and Education (ICFRE) and a biodiversity conservation project in seven national parks in seven different states in coordination with the MOEFCC.

The achievement and outcomes of social forestry programmes have been discussed in the previous chapter. The sector-wide projects had as one of their objectives ensuring support for the National Forest Policy of 1988, with a focus on institutional reforms, management and geographic information systems, upgradation of nursery and plantation technologies, improvement of forest productivity, afforestation and reforestation, and biodiversity conservation. All the projects had peoples' participation as the key approach—popularly called 'joint forest management' (JFM)—to regenerate forests and an eco-development approach at the forest fringes interfacing protected areas (PAs). There is no denying the fact that these projects had a positive impact on development of the forest sector and helped in modernizing forest departments by encouraging new thinking and new technologies, such as improved plant production and plantations and use of information technology like MIS and GIS. The other donor agencies, such as the Japanese, learnt from these projects and avoided pushing for difficult reforms, as they felt that it could be an irritant for bilateral relations and they did not have the kind of leverage the WB enjoyed in those days.

One of the WB-assisted projects, titled the 'India—Ecodevelopment Project' and prepared and appraised in 1995, became a victim of unwanted controversy that would in 1997 see its task manager removed for no fault of hers as the WB management unabashedly blamed her for the controversy. The project design was highly innovative, as it aimed at reducing human pressure on PAs and reducing the negative impact of PAs on human beings living in and around the PA. It aimed to resolve the man-animal conflict through innovative eco-development activities by organizing people in village-level eco-development committees (EDCs) and supporting them with alternative livelihood opportunities, thereby enlisting their cooperation in conserving biodiversity in the adjoining PA.

One grave mistake the WB task manager made was that she thought that stakeholder consultations, including meetings with NGOs, press

(Continued)

(Continued)

briefings and disclosure of information to the public during the process of project preparation and appraisal, would enhance public support and bring laurels for the innovative project design. But it recoiled and what happened was the reverse. The NGOs and even the press became hostile to the project. The main criticism was that the project would deny access to and deprive the tribal and other village communities living in and around the project PAs from meeting their subsistence needs, displace them physically and deny them livelihoods from the forests. They contended that the intentions of the project were to keep people out and result in their economic displacement.

One of the NGOs, the Centre for Science and Environment, even launched a campaign against this project and placed an advertisement in a national daily about the campaign, inviting people to join it. Some WB staff, particularly in the social development sector, also opposed this project. A WB lawyer who was part of the team and even some managers called it a 'tiger versus tribal' project. The internal and external opposition must have caused considerable discomfort for the task manager. The World Bank may not really be so far above reproach as its own personnel claim as it may appear to most outsiders. There have been allegations of professional and personal rivalries.

Though the above project was approved by the WB board in early 1996 and was launched, there were many problems throughout its implementation. The NGOs in Karnataka complained to the WB inspection panel that the project was violating the WB's own operational policies relating to indigenous people and their resettlement and rehabilitation, as there was economic displacement of local tribal and village communities living adjacent to Nagarhole National Park in Karnataka. The WB management's first step was to replace the task manager with a senior member of the social development staff. The inspection panel's probe caused uncomfortable moments for some staff and managers. However, the findings of the panel were not entirely unfavourable, as the project had included adequate safeguards to mitigate any possible adverse social impact and indeed, there was no likelihood of any adverse environmental or social impact—if anything, the impact would be positive.

The worst of it were the internal politics and the insensitivity of the WB management in dealing with this difficult situation. Many international organizations, including the WB, do not recognize and reward hard work, innovation, commitment and sincerity if anything goes wrong for no fault of the staff; on the contrary, it results in their discouragement. Yet the lessons learned from this project design were used and adopted in designing India's WB-funded rural livelihood projects in many states. The eco-development approach to minimize negative impact on people in PAs and reduce negative impact of biodiversity conservation on nearby communities was widely adopted in all states, even by domestically financed programmes, as this is the only way to successfully gain the people's cooperation in the conservation of biodiversity. The approach is being implemented even today, two decades after the WB-funded project was completed.

inspection panel, as they alleged that in Madhya Pradesh the WB was violating its own operational policies relating to indigenous peoples and their resettlement and relocation. The WB's social development wing had contacts with these organizations and claimed that they could fix this issue. The WB sent a team of its social-development staff to investigate, who assumed the role of the inspection panel and, while on a field trip, had an altercation with the forest department officials. No practical solution could be found and, with or without instigation from the WB staff, these organizations organized a demonstration at the WB New Delhi office in 1999, which disturbed the country director and other managers; the task manager, as usual, came into the line of fire, as did the entire forestry programme assisted by the WB.

The decline in WB commitments after 2000 was due to a radical social agenda for promoting tribal and traditional land rights and an aversion to the state forest departments and forest officers. There were already attempts to discredit forest departments because of an allegedly poor human rights record—or so it was argued internally—and resentment against the forest departments' enforcement of draconian laws that deprived poor forest dwellers of their traditional livelihood was already simmering; these came to the surface in 1999. The WB sought to exclude the forest department from controlling forest lands and passing on forest-land ownership to local communities, really a retrograde step borrowed from African and Latin American institutions. The WB sought to establish an analogy with American Indian people here, not acknowledging that the cultures and histories of the two nations and their people had evolved differently.

Internal contradictions involving forest sector finance with confused signals 2003 onward found unrealistic expectation at the WB including:

(a) Continued engagement with the forestry sector without project funding after 2003 and changing the rules of the game implying what follows

(b) An expectation that India would approach the WB, accepting all conditions before funding was provided

(c) A desire to take away jurisdiction over forest lands from the forestry departments and pass it on to the rural-development or tribal-development agencies

(d) An expectation that forest departments would first implement a tribal dwellers' rights act and then the WB would provide funds directly to the tribal communities, excluding the forest department

(e) The WB staff's abhorrence of the forest departments and forest officers of India (resulting in a contemptuous attitude)

(f) Doublespeak by the WB: while appreciating the forestry projects' successes, they abandoned the Jharkhand and Madhya Pradesh projects under preparation in phase II

(g) Some WB staff shared the views of overt and covert Marxist organizations operating in tribal areas

(h) The WB was too far inclined to support community-driven development and livelihood programmes, even though these subsumed forestry interests, to justify that it was supporting poverty reduction.

In 1997, the WB country director first spelled out the position that the forest departments were military-like organizations and were adversarial to the tribal peoples. The 'tribal versus tiger' position was articulated, indicating that if you supported one, the other got hurt. The WB's South Asia management was hostile to foresters and forestry, and one vice president even wrote in her travelogue that when she came across a forester during her first South Asia trip, she wanted to take out her sword.

The country director for India insisted (in 1998–1999) that there should be a strategy for supporting forestry in India. A number of versions were proposed but not accepted despite a clear policy shift at the WB to poverty reduction as the key objective. The sector director of the time supported India's forestry programme, but starting in 2002, new management turned hostile to it and bought advocacy of anti-forest group. By 2003, the WB management had become totally hostile to forestry in India.

In a concession in 2002 to the Andhra Pradesh chief minister, the WB President, James D. Wolfensohn, announced while addressing village forest committees (Vana Samrakshana Samithis) in the Mehboobnagar district of Andhra Pradesh that a repeater forestry project would be given to the state. This caused a furore within the WB and even the WB India country director later questioned why there should be a standalone project in the forestry sector. The country director said, 'We are vulnerable to NGO criticism and we should avoid working in forestry sector in India. It gives us headache.' A senior manager at the WB headquarters said that the WB provided 5 per cent assistance to the forestry sector and got 9 per cent problems. As a consequence, the WB did not approve any forestry project after 2002 and its management yielded to a hostile internal campaign unleashed by the South Asia Social and Rural Development divisions.

Part II

Forest Resource Management

CHAPTER 4

Forest Resources

The forest sector is important as it is the nation's second-largest land use after agriculture. It is a source of goods and services used by all of society. About 70 per cent of India's rural population depends on fuelwood to meet their domestic energy needs. In remote forest-fringe villages, millions of tribal communities and other local people depend on the forests for their subsistence and livelihood. Indirectly, forests protect important catchments for water, conserve soils, ameliorate climate change and combat global warming and desertification. Both wood and non-wood products are important; so are the environmental services of forests and rare species of flora and fauna.

As mentioned in Chapter 1, the systematic management of forests started in India in 1864 with the appointment of the first Inspector General of Forests by the British rulers, the constitution of an Indian Forest Service (IFS) and the promulgation of the Indian Forests Act of 1865, revised in 1878. During the last century, the main task of the IFS was to create reserved and protected forests. In these forests, systematic management and working plans were introduced. In 1927, the new Indian Forest Act was promulgated and in 1935, forests became a provincial subject. The period till 1980 was one of heavy logging for revenue generation for the provincial governments. In the 1980s, timber leases to private contractors were discouraged and phased out. Massive afforestation and social-forestry programmes were undertaken. To reduce pressure on natural forests and meet the demand for fuelwood, fodder and small timber, non-forest lands were planted with trees under the social-forestry programme. The Forest (Conservation) Act (FCA) was promulgated in 1980 to restrict indiscriminate use of forest land for non-forest purposes. A new forest policy was adopted in 1988.

Forest Cover and Degradation

The forests and forest land continue to be under heavy socio-economic and political pressure. Reduction in forest lands and depletion of forest cover go hand in hand. To meet the demand for fuelwood, fodder, poles and timber, the forest cover has been reducing, particularly on the fringes of villages, and forest boundaries are receding. Lush green forests exist only in remote and inaccessible areas, away from human settlements. The indiscriminate felling of trees and grazing in forests cause soil erosion and landslides in hilly areas, prevent regeneration of forests, alter biogeochemical cycles of ecosystems and degrade the land. The high intensity of degradation makes recovery or restoration tasks very difficult and expensive.

The degradation is not only indicated by decline in crown density but also soil erosion, absence of humus and organic matter, lack of natural regeneration, absence or lack of ground flora and shrubs, and poor productivity. Nevertheless, it is claimed by government agencies that the trend of fast degradation has been arrested and, with the restoration of degraded forests through massive joint forest-management activities, the cover is improving. The main area of concern is that 40 per cent of forest cover is open forests with a crown density between 10 and 40 per cent.

India's per capita forest area is 0.08 hectares (it was 0.20 hectares in 1951), which is one of the lowest in the world (the world average is 0.64 hectares). India has one sixth of the world's population. North-East India, the Himalayan states, Madhya Pradesh, Odisha and the Andaman and Nicobar Islands are rich in forest resources (26–89% of land under forests). Punjab, Haryana, Gujarat, Rajasthan and Uttar Pradesh plains are poor in forests. There is diversity of forest composition, which vary from the mangrove and littoral swamp forests to temperate and alpine forests, the maximum area being under tropical dry and moist deciduous types of forests.

Between 1950 and 1980, India lost 4.3 million hectares of forest land due its diversion to non-forest uses (agriculture got the largest share—2.6 million hectares) at an average rate of 150,000 hectares per annum. With the promulgation of the FCA, that rate has come down to about 50,000 hectares per year. About 3 million hectares of forest lands are under encroachment. About 50 per cent of the country's total livestock graze freely in the forests. About 200 million cubic metres of fuelwood is gathered every year by the local communities. About 8 million people are practising shifting cultivation in forests, mainly in the North-East, Odisha and Andhra Pradesh. Thousands of hectares of forests are damaged annually by incendiary fires.

Forest Productivity

The National Forest Policy of 1988 had an objective of improving forest productivity to meet essential national needs. This was envisaged as being achieved through the application of scientific and technical inputs. The forestry programmes were to be oriented to narrow the gap between demand and supply of fuelwood through afforestation, while avoiding clear felling in well-stocked natural forests and discouraging introduction of exotic species. However, the productivity of forests has not improved but rather has decreased due to rampant ongoing degradation all over the country.

In 2015, the total growing stock, as estimated by the FSI, was 4,218 million cubic metres in the government-owned forests and it was 1,604 million cubic metres outside state-controlled forest areas (trees outside forests) (FSI 2017). Outside the recorded forest area, the tree cover is over 9.3815 million hectares, which has been included in the total forest cover in the country of 70.633 million hectares. These are mostly tree patches of less than one hectare each. 'Trees outside forests' are all trees, irrespective of the size of the patch, outside the forest area de jure controlled by state governments through their forest departments. The average productivity outside the government-managed forests is more.

The Ministry of Environment, Forest and Climate Change (MOEFCC) recognized that timber production from the government forests has declined due to restrictions on cutting in many states and the direction of the Supreme Court of India, as well as the increased emphasis on biodiversity conservation (ICFRE 2010). By 1998, the total annual timber production from the government forests had declined to 2 million cubic metres. Since 2005, the production has ranged from 2.18 million cubic metres to 2.6 million cubic metres annually. In contrast, the annual timber production from outside the forests was estimated to be about 44.3 million cubic metres.

Import of Wood, Pulp and Paper

To meet the huge demand for timber, there is huge dependence on imported timber. The imports have been increasing every year due to increased demand and lack of indigenous production. For example, in 1992, 867,000 cubic metres of timber was imported, and in 2010, it was 5,909,360 cubic metres—that is, about seven times the 1992 figure. Imports from Malaysia, Myanmar, New Zealand, Ghana, the Ivory Coast and Gabon form the bulk of these, and 15 per cent of the timber is that of teak species alone.

Although 77 million hectares are recorded as forest area and India has vast areas under sal and teak forests, the country is not self-sufficient for wood and wood products. Most forests are so badly degraded that any harvesting without simultaneous reforestation will make the land totally barren and probably accelerate the process of desertification in sub-humid areas. The average growing stock of 54 cubic metres per hectare in India's forests is very poor, and the lowest among the world forests. One reason is that more than 50 per cent of the forests have been degraded due to heavy logging and destruction over the last hundred years and more. Forest productivity remains low and has not improved perceptibly. The greatest challenge for India is to restore forest cover and productivity over the next fifty years. The forest cover has been categorized by the FSI as given in Table 4.1.

This shows that open forests and scrubs constitute 46 per cent of the forest areas where crown density is less than 40 per cent. Only 13

Table 4.1 Forest Cover Categories (by canopy density)

Forest condition	Density range (percent-age)	Million hectares	Percentage	Assumed average density (percentage)	Weighted average reduced to 100% density (area in million hectares)
Very dense forests	70–100	9.8	13	85	8.33
Moderately dense forests	40–70	30.8	41	55	16.94
Open forests	10–40	30.2	40	25	7.55
Scrub	0–10	4.6	6	5	0.23
Total		75.4	100%		33.05

Source: FSI (2017).

per cent of the forests are dense, with a crown density above 70 per cent. Moderately dense forests have a crown density of 40 to 69 per cent. The degraded forests are at 46 per cent and forests with more than 40 per cent density are 54 per cent of the total area. In other words, 89 per cent of forests are poorly stocked. It can be assumed that 60 per cent of the forests have only a 25 per cent potential stocking of trees. These are highly degraded. It can also be assumed that not more than 10 per cent of the forests are fully stocked.

The last column in Table 4.1, where forest area has been reduced to full stocking or 100 per cent crown density by applying an average density factor (mid-point of the range), reveals that the average productivity of public forests is about 33 per cent of their productive potential. This also gives an overview of the vast land resources being underutilized. This means that, virtually or notionally, the equivalent of 33 million hectares of forests are fully stocked and virtual blanks cover about 70 per cent of the forest lands. However, these assumptions are questionable because the forests also have natural blanks—bare rocks, waterbodies and grasslands. It can safely be construed that forest density and productivity can be increased to three or four times the present status, provided the required financial, technical and institutional inputs are made available.

During the last three decades of forest-policy implementation, dense forest cover did not improve or significantly increase, though open forests increased. One evidence of low productivity is that timber imports have gradually increased. The forests, which cover 23 per cent of the country's land area, have failed to meet the demand from people for timber, pulp and other forest products. Remarkably, the output from 2.5 million hectares of private land is more than that coming from the state-owned forests. Yet the government is occupying 77 million hectares of forest land—that is, about one fourth of the country's total land mass—with its huge infrastructure and Forest Service, and yet failing to provide optimum output from the forests.

Carbon stocks in the forests are currently estimated at 7,044 million tonnes. If one tonne of carbon sequestration fetches even US$10, there is a possibility that US$159.640 billion could be earned for sequestration of 15,964 million tonnes of carbon dioxide over a period when forests are brought to the full stocking. The country needs an investment of US$50 billion[1] for forest restoration (equivalent to 50 million hectares of reforestation or afforestation) over a period of 30 years. This is ₹10,000 crore per year. For each additional tonne of carbon sequestration, US$3.13 is needed.

The facts discussed above indicate that the 1988 Forest Policy to a large extent failed in its objective of improving productivity. The government forest-management agencies should be forthright in acknowledging this fact and avoid the temptation of being happy that the trees outside the forests have good growing stock and are meeting demand from the wood (mainly pulpwood), veneer and plywood industries. That is purely a private sector, in which the market-driven forces have encouraged the landowners to plant trees to earn additional income. The poor productivity in state forests also reflects an administrative failure in that forestry establishments had been mining

---

[1] Per hectare average cost of afforestation/reforestation being US$1,000 at the 2017 price level.

timber till the Supreme Court of India intervened, but the damage had already been done.

Forest Cover Assessment

In the State of Forest reports, the latest figure for forest cover is compared with that from the previous report. However, any direct comparison between 1999 and 2001 reports would be invalid because of different techniques and scales used. According to the State of Forest Report (FSI 2011) of the Forest Survey of India (FSI), there was an increase of 1.0100 million hectares of dense forest cover and 3.7890 million hectares of open forest. There has also been a continuous decrease in scrub forests, which have canopy density less than 10 per cent (Bhojvaid et al. 2013). The net difference in forest cover per the 2001 and 1999 assessments shown in Table 4.2 is composed of two entities: (a) a difference due to technical factors and (b) the real change in forest cover during the intervening period. However, it is not practically possible to determine the real change in forest cover during the period between the 1999 and 2001 assessments (FSI 2001). However, Ravindranath et al. (2012) claim that there has been no actual increase in forest cover. They based their findings on the FSI's own publications. The assertion of increased forest cover appears to be unreliable, as has been explained by the State of Forest report (FSI 2009), where it has clearly been stated that due to significant technological and methodological changes, especially after 1999, the forest cover area data generated during subsequent assessments by the FSI were not comparable. Thus, it would be irrelevant to compare the 1991 assessment with that of 2011 and assume an increase in forest cover (of more than 40 per cent crown density).

The decadal change in forest cover from 1997 to 2007 was only 31,349 square kilometres (refer to Table 4.2). The biennial reports of the FSI have mostly been showing an increase in the forest cover. For example, between the 2001 and 2015 assessment, it has reported an increase of 2.18 million hectares in forest cover (Table 4.2). However, the results of the FSI survey may be questionable, though this remains the only data to rely upon. The satellite data interpretation keeps

Table 4.2 Forest Cover in India (in sq. km)

Category	1999	2001	2003	2005	2005 (revised)	2007	2009	2011	2013	2015	2017
Dense forests (crown density > 40%; 40–70% from 2005)	377,358	416,809	339,279	332,647	319,948	319,012	320,238	320,736	318,745	315,374	308,318
Crown density above 70%			51,285	54,569	83,472	83,510	83,248	83,471	83,502	85,904	98,158
Open forests (crown density 10–40%)	255,064	258,729	287,769	289,872	286,751	288,377	288,728	287,820	295,651	300,395	301,797
Mangroves	4,871	Included in above									
Total forest cover (percentage of geographical area)	637,293 (19.39)	675,538 (20.55)	678,333	677,088		690,899		692,027 (21.05)	697,898 (21.23)	701,673	708,273 (21.54)
Scrub	51,896	47,318	40,269	38,475		41,525	42,050	42,177	41,383	41,362	45,979
Total recorded forest (percentage of geographical area)	765,253 (23.28)	768,436 (23.38)						734,204	739,271 (22.49)	744,035	754,252 (22.94)

Source: FSI (1999, 2001–2017).

changing, besides the technological changes in remote sensing. Adjustments are always made in interpretations to account for such changes. From 2013, better satellite imagery, with high resolution, has become available, which is improving the interpretation and analysis. Thus, when one looks at the forest cover reported by the FSI in 2007 and 2015, there is a sudden jump of about 1.08 million hectares in forest cover. On the other hand, deforestation is also taking place all over the country. On an average, about 50,000 hectares have been cleared under official orders for non-forestry activities, besides unauthorized conversion of forest land to other uses by local villagers. Another interesting finding of the FSI is that during the last three decades, open forest area (with canopy density below 40 per cent) is consistently increasing, pointing to the fact that forest degradation is progressing at an alarming rate.

The biennial reports of the FSI that bring out the data on monitoring of forest cover and tree cover in India do not really show a tenable, significant increase in the forest cover since the beginning of the 1988 Forest Policy. Interestingly, tree plantations were raised over an area of 32.48 million hectares between 1986 and 2007, and surprisingly, this amazing figure has not been captured through satellite imageries as interpreted by the FSI. If one adds to this figure the tree plantations from 2007–2017 (a period of 10 years), the achievement of plantations will be mind-boggling. Even if one assumes (in the absence of available data) that about 1.5 million hectares were planted every year, the last 10 years would have added 15 million hectares, making the progress about 50 million hectares.

Demand and Supply of Forest Products

In innumerable locations, people living in the vicinity of forests are buying imported timber for house construction due to non-availability of local timber as well as high prices. This not only results in a sense of deprivation, but also contempt and disrespect for the public institutions. It is also assumed that the demand from local communities for fuelwood, fodder and small timber has been increasing with population increase, poverty, lack of alternatives and depletion of forest resources. No reliable information is available on the production and consumption of wood. According to one estimate, the production trends have been as shown in Table 4.3.

A study conducted by the MOEFCC in 1999 made projections of the demand for industrial wood, which is shown in Table 4.4.

India's population is over 1.3 billion and is still growing at the rate of about 1.6 per cent per annum. Pressure on land will continue to increase and intensify. More forest products for consumption and more forest land for agriculture and development projects for infrastructure, mining of ores and minerals, and urbanization will be required. The huge demand and supply gap, bridged today by importing huge quantities of wood and pulp that deplete the country's foreign exchange, will inevitably increase. The demand from a

Table 4.3 Wood Demand and Supply (in million m³)

Consumption/output	2001	2011
Industry, furniture, agriculture (utilization)	82	91
Output from state-owned forests	12	3
Output from privately owned plantations, agroforestry	53	43
Deficit	14	45

Source: Bhojvaid et al. (2013).

Table 4.4 Wood Demand Projections (million m³) Supply/production projections from all sources (million m³ in round wood equivalent)

Wood type	2000	2005	2010	2015	2020
Pulpwood	8.76	14.32	21.92	34.67	45.86
Sawn wood	23.21	28.97	36.75	45.44	54.66
Panel wood	17.80	21.39	25.70	30.96	38.13
Other wood	8.00	9.03	10.50	12.09	14.25
Total	57.77	73.71	95.07	123.16	152.90

Source: MOEF (1999).

growing population has put an enormous pressure on the forests. It cannot be met on a sustainable basis, though the supplies are being supplemented by illegal removal of forest products and imports, and also from non-forest sources.

According to one estimate (MOEF 1999c), the value of goods taken from the forests of India is US$43.8 billion, or about ₹180,000 crore at the 1998 exchange rate. Other estimates show higher or lower figures, one being total removals from the forest of about US$7.1 billion, or ₹30,000 crore. This does not include the value of environmental services provided by the forests, which according to one estimate are equivalent to US$19 billion per year.

Non-wood or Non-timber Forest Products

A comprehensive survey of non-wood or non-timber forest-product (NWFP or NTFP) resources has not been carried out in any state of India. Vast quantities of NTFP are locally collected and traded. There is a variety of such products and it is difficult to prepare an inventory. However, a scientific management system is not in place for managing, harvesting and marketing of these products. It is claimed that the growing stock of NTFP is increasing with the progress of joint forest management (JFM), but it may decline if overexploited. With free access for NTFP collection by local communities, which has removed regulatory control of forest areas particularly in tribal areas, NTFP resources will be depleting fast and some species may even become extinct.

Marketing of Forest Produce

State governments exercise direct control over exploitation and marketing of forest produce from the public forests. Concessional pricing of produce from public forests and import of wood limit the incentives to develop private agroforestry. India's market for forest products is essentially domestic. The situation of the forest sector is one of market failure in India, in which economic efficiency has not been achieved through following a market mechanism (NFAP 1999).

Endangered Biodiversity

India's biodiversity is rich, unique and endangered. India is one of the 12 megadiverse countries in the world, which collectively account for 60–70 per cent of the world's biodiversity. Its 10 biogeographic regions represent a broad range of ecosystems. India has 6 per cent of the world's flowering plant species and 14 per cent of the world's avian fauna (World Bank 1996). As many as 3,000–4,000 plant species are endangered. This indicates a widespread degradation of ecosystems and habitat that has a serious economic implication. From only 10 national parks and 127 wildlife sanctuaries in 1970, the protected area (PA) system now covers 80 national parks and 441 sanctuaries. The PAs cover 14.8 million hectares, which is 4.5 per cent of the country's land area and 14 per cent of the forest area.

Contribution to GDP

According to one estimate, wood, grazing, fodder, medicinal plants and non-wood construction material account for US$43,843 million per year. This does not include a number of NTFPs extracted from the forests (UN-CSD/IPF cited in NFAP 1999).

A serious limitation of this data is that it represents only the officially recorded forest production, and a substantial quantity of production goes unrecorded. This unrecorded production comprises (a) authorized (but unrecorded) and unauthorized removals of timber and non-timber forest produce (e.g., fuelwood) and (b) unrecorded production from private and non-forest lands. Other limitations of the data include time lag, incomplete coverage, non-availability of species-wise production of major forest produce and non-availability or poor quality of data on the production and prices of NTFP.

The estimates are that forestry contributes less than 2 per cent to the gross domestic product (GDP), though this does not take into account total products and thus is underestimated. The contribution of forestry to the GDP has been declining over the years (2.9 per cent in 1985 to 1.7 per cent in 1990 and 1.3 per cent in 1995) because of reduced harvesting of forest produce from natural forests, poor productivity of natural forests and plantations, conservation policies being pursued by the government and heavy reliance on import of wood, pulp and other finished forest products. This has also resulted in declining income to the state from the forestry sector after the mid-1980s. According to Chopra, Bhattacharya and Kumar (2001), the estimated value of goods and services provided by the forestry sector is ₹258.94 billion, the GDP from it is ₹230.03 billion and the sector contributed about 2.37 per cent to the GDP.

Investment in Forest Development

State governments do not allocate adequate budgets for the management of forest resources. The forestry sector annually receives hardly 1 per cent of the total development plan allocation, though the removals amount to ₹300 billion. The forestry sector gets about ₹20,000 million annually (equivalent to US$350 million), which is distributed to all the states. A significant part of this allocation comes from donors such as the government of Japan (via the Japan International Cooperation Agency, or JICA).

Forest Protection

Forest protection using the traditional approach has become an arduous task. Forest protection was organized in the nineteenth century with an unarmed forest guard allotted 200 to 2,000 hectares of forests to protect against illegal tree cutting and other activities. This system still continues. However, some states have set up flying squads, mobile squads or forest stations which can rapidly move to the site of a forest offence, and in many states, the support of the local police is sought to deal with a serious situation. Despite all efforts, illegal tree cutting and forest-land encroachment continue. The traditional system of forest policing is totally obsolete. New initiatives with flying squads, forest police, a forest protection force and even arming of field forestry staff have provided some deterrence, but the open resource always remains under threat. Without state forest departments' (SFDs) efforts and hard work, an anarchic situation would have prevailed and forests vanished long back. Vested interests have always sought to undermine the powers and authority of forest staff to placate popular political constituencies. There is a conflict of interest as there is an ostensible commitment to conservation but a focus on short-term, populist gains. Indiscriminate grazing in forests prevents regeneration and degrades site. Though the primary management function of the forest is not to provide fodder for cattle, the freely moving herds of millions of domestic animals depend on the forests for their sustenance and, in the process, overuse and destroy the forest ecosystems. There is a scarcity of fodder and the cattle population is underfed and consequently low-yielding.

CHAPTER 5

Forest Management

The serious challenge now is how to arrest further degradation of forests and how to restore degraded forest ecosystems. Systematic forest management started in India in the mid-nineteenth century during the British colonial rule. The forests as we see them today are a result of more than 150 years of management intervention. Despite huge pressure on forests from heavy logging for revenue generation, unregulated grazing, fuelwood extraction and timber thefts, the state forest departments (SFDs) have been able to save 70 million hectares of forest land, from complete conversion to non-forest land uses, as well as a huge volume of growing stock of trees, other vegetation and wildlife. However, arresting further degradation and restoring ecosystems remain serious concerns, as they need to proceed at a more rapid speed to sustain forest resources and values for future generations.

Not more than 12–15 per cent of forests are well stocked. The Forest Survey of India (FSI), an agency of the Government of India (GOI), carries out monitoring of forest cover in India every two years. Crown density is used as the basis for classification of forest cover in designated forests (state owned).

The vast majority of forests areas are characterized by:

(a) Overall poor stocking
(b) Absence or lack of adequate natural regeneration

(c) Scanty undergrowth, shrubs and soil cover
(d) Absence of fertile topsoil and organic matter
(e) Eroded soil, with sheet erosion, rill erosion and even gully erosion in many localities
(f) Low proportion of younger trees
(g) Altered species composition from that which existed prior to active management
(h) Heavy grazing by local and migratory cattle
(i) Failure or low survival of plantations
(j) Monoculture plantations, vulnerable to pest and pathogen attack
(k) Depleted biodiversity

In India, forests are not being managed on a sustainable basis. It has been impossible to maintain the ecological integrity of forests for the following reasons:

(a) Past management practices focused on a single objective of wood harvesting.
(b) Forest areas that were heavily logged over in the past did not regenerate, though it was assumed that forest regrowth would take place naturally.
(c) About 89 per cent of forest ecosystems are in varying states of degradation.
(d) Heavy grazing by livestock in forests indicates that legally designated forests have pasturing or ranging as the de facto primary land use and forestry is a secondary land use.
(e) Indiscriminate fuelwood extraction by rural communities disturbs forest health and ecological integrity.
(f) Forest boundaries are not legally sacrosanct. Encroachment on forest land for agriculture is common.
(g) Leaf litter on forest floors is collected for domestic energy or manure.
(h) Lopping is done for fodder and fuelwood as well as for green manure or composting material.
(i) Human-induced forest fires are common.
(j) Forest restoration through reforestation and afforestation so far has not been able to offset deforestation and forest degradation.

The objective given priority was the yield of selected timber-producing species. The yield was not possible to sustain in perpetuity. Forest health and well-being were implicitly included as objectives of forest management through a working plan, emphasizing sustainable yield management along with conserving soil, fauna and flora. However, these other objectives were not given adequate attention in the implementation of management plans. The yield was neither sustainable nor possible to maintain in perpetuity. While managing forests for timber harvesting, attention was not paid to dependence of communities and other land use of forests like livestock grazing, domestic energy and livelihoods (Khan 1987). The local population was used as a labour force and not a stakeholder. This was a legacy of the colonial perspective that the land belonged to the ruler who could do with it as they pleased. Concessions or rights were treated as a reward for servitude. The feudal system that perpetuated poverty, hunger and slavery disenfranchised millions who dwelled in the forests and on their fringes from their age-old rights and privileges over the nearby forests. However, the increasing population and unlimited demand and urge for consumption of forest-based resources, in any case, could have destroyed all forests beyond recovery.

Impact of Past Management

The classical forest management philosophy did not take into account ecology, multiple-use forestry, rural livelihoods and a long-term vision. The colonial rulers carved out forests from wastelands, and the rulers of princely states claimed control over all land that was forested but not owned by anyone. While doing so, though, the British-ruled provinces undertook a process of settlement where only the rights of settled villagers were inquired into and not those of landless nomadic tribes who depended on hunting and foraging. The objective of the imperial administration was to mine timber for shipbuilding and later for laying a railway network in India. The natural forests were a good source of commercially valuable timber species. Thus, the main objective of forest management was to harvest timber and also convert mixed-species tree stands to a single-species, even-aged forest. The basic management principles were brought over from Franco-German forestry practices and the silviculture and management systems introduced in India heavily relied upon assumptions that any logging would be followed by natural regeneration and that the yield of timber would be available in perpetuity.

A planning system was developed for producing working plans for a 10-year cycle (initially, more than 10 years too) to assess growing stocks and estimate allowable annual cuts (yields) of timber. This was also reflected in forestry training, in that the principles of agronomy were borrowed and the forests were termed 'crops'. Forestry personnel were trained in the art of logging, growing forest crops, developing forest industries, building infrastructure (road, culverts, small bridges and small buildings) in remote forest areas, and protection of forests from timber thieves, pests, insects and pathogens. This philosophy was also reflected in the forest laws and policies, starting from 1892. At that time, the human population was not much, and forests and lands were assumed to be abundant and to be used as thoroughly as possible. The management adopted the following classical (silvicultural) systems:

(a) The uniform system (or its variant), seed-tree system or shelterwood system
(b) Selection system or group selection system
(c) Clear-cutting with artificial regeneration
(d) Coppice system, coppice with reserve or coppice with standards

The objective was to harvest timber on a sustained yield principle and achieve natural regeneration or plantations in clear-cut areas. A certain number of seed trees were retained after major felling, and commercially valuable species were favoured, with the removal of other tree species, in the shelterwood system. In the selection system, trees with the minimum exploitable diameter were harvested. Thus, it involved the liquidation of trees of higher diameter classes belonging to commercially marketable species (for example, Himalayan pine forests). Clear-cut areas sooner or later became barren due to lack of protection and poor maintenance of plantations with the required inputs. All these operations, coupled with heavy and unregulated fuelwood extraction and grazing, resulted in degradation and depletion of forest resources across the country.

The main role of the state forest departments (SFDs) was to generate the maximum and increasing revenue for the state. After the country attained independence in 1947, colonial policies were still continued and cash-strapped states demanded more and more revenue from the forest sector. The main role of the SFDs was still to meet the state finance deportment's revenue target. The financial resources for the development of other sectors in most states came from the forests, without a corresponding and much-needed investment for the improvement of forest resources.

National Forest Policy: The Management Approach

The objective of reversing or even halting further deterioration of forests poses a serious challenge to not only the practitioners of forest management but for the country as a whole. As a collective national conscience, this generation is failing (as the previous generation did) at leaving behind a healthy natural resource for future generations. Sometimes democratic institutions guarantee a healthy natural environment and natural resources (though only in the long run) and sometimes, with poor governance, they become instrumental in destruction of resources instead. Under various orders from the Supreme Court of India, there is a ban on cutting of green trees in the Himalayan region. The court directives have also imposed a ban on harvesting of timber without ensuring regeneration or replanting as well as harvesting without a valid working plan for the relevant forest division. Harvesting of plantations, however, is permissible. The Supreme Court's directives were issued from time to time as the Executive failed to implement its own policies. However, on flimsy justification, such as removal of 'dry', 'damaged', 'dead' and 'dying' trees, which are allowed to be harvested, even some green trees may be cut down. Green trees continue to be cut down illegally, with or without the collusion of SFD staff. Vested interests do not desist from diluting the regulatory framework and sometimes even judicial directives.

The National Forest Policy of 1952 could not be successfully implemented. This is evident from the preamble to the National Forest Policy of 1988 (the current policy), which acknowledges that

> Over the years, forests in the country have suffered serious depletion. This is attributable to relentless pressures arising from ever-increasing demand for fuel-wood, fodder and timber; inadequacy of protection measures; diversion of forest lands to non-forest uses without ensuring compensatory afforestation and essential environmental safeguards; and the tendency to look upon forests as revenue earning resource.

The Forest Policy ideals have not been adequately supported by actions on the ground. First, the policy reiterates the goal of 33 per cent of the country's land being put under forest cover. This aspirational goal remains a distant dream. The current forest policy gives supremacy to environmental considerations in forest management, and all other objectives have been made subservient to it. Yet how far the ecological goals have been achieved is a matter of debate. Second, the policy stipulates that local people have first charge of forests and that their bona fide needs from the forests should be satisfied. These ideals have not been adequately backed by actions either. The policy stipulations relating to grazing rights and concessions have not been successfully implemented and it has been business as usual. Also, the goal to enhance forest productivity has, by and large, not been achieved as is evident from the successive reports of the FSI. Conversion of forests to non-forest uses continues, with increasing demand for forest clearance for mining and infrastructure development, which has lately involved diluting the forest clearance process by allowing in-principle approval for starting up projects on forest land, to be formalized only later.

Forest Ecosystem Loss and Degradation

Despite the conservation-centred policy and massive afforestation and reforestation efforts, forest degradation could not be arrested or reversed. The main causes of forest ecosystem loss and degradation (consequent to increase in population and rising socio-economic expectations of society) are listed below:

Deforestation

(a) Conversion of forest land to agricultural land
(b) Clearance for urbanization
(c) Clearance for infrastructure development—roads, railway lines, dams, transmission lines, irrigation canals, water reservoirs, and so on
(d) Mining of ores and minerals
(e) Quarrying of stones
(f) Settlement of displaced persons
(g) Government buildings
(h) Industries

Degradation

(a) Heavy exploitation of commercially important species
(b) Commercially oriented forest management involving clearance of mixed forests and their replacement by a single species
(c) Heavy reliance on (anticipated) natural regeneration as a management strategy
(d) Low investment in regeneration or forest restoration after harvesting
(e) Human population growth resulting in increased demand for forest products
(f) Tragedy of the commons
(g) Poverty forcing people's heavy dependence on
  (i) Wood as domestic energy source
  (ii) Wood extraction for fuelwood or small cottage-based industries such as tea-leaf curing or tobacco curing
  (iii) Free and unregulated livestock grazing
  (iv) Unsustainable and destructive selective harvesting of non-timber forest products (NTFP)

   (v)   Shifting cultivation
(h)   Forest fires, almost all human-caused

Policy and market distortions

Policies of distributing timber free or at concessional prices involved increasing demand and created political pressures to meet these demands by excessive harvesting. NTFP collection, barring a few species, was allowed to proceed unregulated and free of cost, which has not only resulted in unsustainable and destructive harvesting, but also the disappearance of many species in many areas.

Intersectoral or cross-sectoral policies

Policies of other sectors have had and continue to have serious adverse impacts on forests. For example, the policy of the livestock sector that promotes increase in population of sheep, goats, cows and other cattle puts increased pressure of grazing on forests, without appreciating the carrying capacity or availability of fodder and pastures. Mining, infrastructure and agriculture are other major sectors that involve forest clearance. Agricultural and tribal policies in certain aspects also conflict with the forest conservation objectives.

Institutional capacity

Policies and laws have been implemented half-heartedly, partly due to lack of resources and partly due lack of political will, resulting in the absence of desired outcomes. The inadequate organizational capacity of SFDs is also detrimental to effective management of forest resources. During the last two decades, about 30–35 per cent of the staff positions have remained vacant and, simultaneously, responsibilities have multiplied manifold. Inadequate institutional capacity limits SFDs from reaching the desired level of functioning and adversely impacts sectoral performance, thereby creating stress and negative public perception.

Land tenure

Incomplete settlement process, clear demarcation of legally designated forests' boundaries and entries of forest land in land revenue departments' records resulting in disputed land-tenure, and partial implementation of the Forest Rights Act (FRA) have been affecting forest management. Land tenure in many areas is still not settled, which allows illegal occupation of forest land and change of land use.

Political support and alliances

One discouraging fact is that both central and state governments have always given low priority to the forest sector in the socio-economic development agenda, except in the decade of the 1980s, when the conservation movement got the support of the central government. During this period, substantial funds were made available for forest development and conservation, a Forest (Conservation) Act (FCA) and a new Forest Policy were promulgated, and people's participation in afforestation and reforestation promoted. However, as of now, the forest sector neither enjoys any noticeable political support or clout, nor is it in alliance with the media, civil society, politicians, bureaucrats or the public at large. This isolation and even an apparent contempt for forest departments has been a serious constraint for the last six decades and more, having a detrimental influence on forest management and demoralizing forest personnel.

Forest productivity has declined over the years; so has harvesting of forest products. The current management strategy for forests is a mixed one, focused on conservation and harvesting of forest products on a supposedly sustained yield basis. Wood harvesting has declined over the years. The bulk of timber, pulp and even finished furniture used in India is imported from other tropical countries, such as Malaysia and Indonesia. The condition of most forests, which have been under regular management for a long time, is that they are heavily degraded and have mostly middle-aged and a few mature trees, with little or no regeneration, an absence of young poles and an understorey, eroded soils devoid of humus and organic matter, and dry land with no moisture content (most of the year) in the soil. Probably these forests have been harvested repeatedly at short intervals without allowing the system to reach full maturity and ensure nutrient cycling. All this makes these areas highly vulnerable, and they are slowly turning arid; it is just a matter of time before these lands become desert. The forests are exposed to unregulated heavy grazing, grass cutting and fodder lopping. Leaves are swept away from forest floors in many areas, impoverishing the site in organic matter and soil nutrients. Topsoil is washed away during the rains due to the soil surface being unprotected by vegetative cover. Biodiversity in the forests is declining and, except a few, most protected areas (PAs) are not being managed actively due to lack of funds and staff. Also, the PAs are not being managed as ecosystems but are being preserved for a particular mammal or bird species, for example, tiger, elephant, bison, rhino or lion.

Classical forest management has not changed over the years, nor shifted to an integrated and holistic management system based on sound ecological principles. Other than expressing a wish to have sustainable management of forests, no serious efforts have been made to adopt new approaches to and technology for forest management. The current national forest policy does not mention sustainable forest management, as this concept was discussed in the forestry sector only after the 1992 Rio Conference. Most Indian forestry professionals equate sustained yield management with the principle of sustainable forest management. The latest National Working Plan Code (2014) also advocates management of forests on a sustainable basis, but fails to explain how? The draft of a new forest policy also includes sustainable forest management as an objective, with inherent contradictions. Forests are not managed as ecosystems where all elements and their functions and biogeochemical processes are allowed to continue to maintain the ecosystem's functionality and integrity. The focus continues to be only on trees but not on shrubs, climbers, herbs, grasses, fauna (and microfauna) and flora, soil, soil nutrients, soil moisture and the fringe human habitations that affect the ecological processes and are critical to ecosystem management.

Two new demands on forests that have emerged recently are biodiversity conservation and climate-change mitigation (carbon sequestration). We do not know what may be the new demands on forests in the future. This requires that forest ecosystems should not be disturbed beyond recovery and their integrity be maintained while using their services on a sustainable basis for the benefit of society.

Worldwide, forest management practices are currently undergoing the most thoughtful, intense and swift changes since their beginning in the nineteenth century. The shift from the principle of sustained yield management of quite a limited number of marketable tree species to the sustainable management of forest ecosystems is transforming what were historically some of the basic forest management principles. The approach to looking at forests for trees or wood is no longer valid today, when it has been established that forests provide important ecosystem services and any interference with the biogeochemical processes in a forest ecosystem and alteration of its structure and functions destroys the ecological integrity, with serious repercussions on the abilities of these ecosystems to provide goods and services to present and future generations of human societies.

Sustainability is now the overarching goal of forest management, which is based on the principle that for present and future generations, the forests should be able to deliver ecosystem services. This requires protection and management of forests as ecosystems and landscapes.

The management of forests to ensure they remain 'closer to nature' has increased significantly in recent decades. The trends of 'nature-based silviculture' or 'close-to-nature forest management' approach in Europe and 'ecosystem management' and 'adaptive management' in North America aim at improving current forest management practices so that they are still profitable, but in a manner more environmentally sound and with more sensitivity to the complexities of nature conservation and the multiple, varying and steadily increasing demands of society, by mimicking natural forest structures, their processes as well as their dynamics. The management of forests in a manner 'closer to nature' is simultaneously accompanied by ever more reliable and refined models, promoting its efficient implementation. The basic idea is to reach a better balance among productive, protective and social functions. Other important goals are to increase economic competitiveness by cost reduction and to increase robustness to withstand climate change (Larsen 2012).

The philosophy of 'close-to-nature' forestry or 'nature-based forestry' emerged in Europe as early as the nineteenth century, through the writings of eminent foresters. Clear-cutting and replanting with a monoculture of conifer species was the most prevalent management practice in Europe, which started in the early nineteenth century. This management practice resulted in site degradation, frequent windbreaks and outbreaks of pest and disease attacks that caused huge damage to spruce plantations in Switzerland. Switzerland shifted to close-to-nature forestry in the late nineteenth and early twentieth century (Larsen 2012). During this period, India was introducing the traditional European forestry practices such as the uniform or regular shelterwood system for changing irregular forests into regular ones, the selection system and its variants and the coppice system to selectively cut down marketable tree species.

The nature-based forestry movement began in Germany in the 1920s. In 1950, a close-to-nature forestry group was organized in Germany, and the group members were foresters and forest owners. The expansion of this movement beyond Germany gave birth to Pro Silva in Slovenia in 1989. Currently, 26 European countries are its members, and the United States of America (USA) and Canada have also joined recently. The members are foresters, forest owners, students and other interested parties.

Obsolete Management in India

It is obvious from the state of forests in India that the classical or traditional forest-management approach has become scientifically obsolete and it has been having deleterious effects on the remaining, logged-over forests, resulting in a serious degree of ecosystem degradation. Therefore, there is a need to shift from the classical management system to a more scientific, integrated and ecosystem-based forest management, with the objective to restore forest ecosystems.

India has vast variations in ecological conditions, including soil, climate and vegetation type. All planning and application of the principles and operations of forest management should therefore be site-specific. Different approaches and methods will be needed for different forest types and varying site conditions in terms of gradient, soil fertility, erosion status, precipitation, soil moisture and other environmental factors, balancing local peoples' dependence on the forests and the need for the forests to have protection from grazing. There is also a need to develop site-specific plans and model management plans on a pilot basis, to demonstrate that ecosystem-based management can be introduced without drastic changes in management procedures and policies.

The methods and techniques to be applied to change forest structure and composition to bring it as far as possible to its original state require introducing a newer version of the selection system for the continuation of forest cover across all forests in a way that ecosystem integrity is restored and the flow of ecosystem services continues on a sustainable basis. When afforestation, reforestation or enrichment planting in degraded forests is undertaken or unsustainable coppice forests are changed to healthy and mixed forests with local species, the principles of Pro Silva as applicable to Indian conditions should be followed.

The necessary paradigm shift will require acceptance of the new principles by the government, foresters and society. Forests are owned by the states, which manage these through their SFDs. The traditional system of working plans governs the management of forests and rarely is any deviation allowed, though not all prescriptions are followed other than the allowable annual cutting. The new concept of management will have to be incorporated in the working plans under preparation and those prepared in future. In addition, the Ministry of Environment, Forest and Climate Change (MOEFCC) and the SFDs can issue guidelines for the management of forests. It will be a challenging mission, as some people will perceive it as a threat to their long-held views, beliefs and conviction in classical management. This is but inevitable and there will be few early converts, but as a campaign is launched and sustained, it will draw wide support.

People at the Centre of Ecosystem Management

A new approach will be desirable to involve local community as well as civil society. The joint forest management (JFM) approach could not be sustained and did not have significant positive impacts on forest resources, barring a few exceptions. It has been a project- and fund-driven activity, which became dormant when both these were withdrawn, reverting to business as usual. The one extreme effort to hand over management to local communities did not succeed, nor another that expected people would protect forests in exchange for a promised share in the forest produce or revenues. An integrated management is much more complex than JFM, as the target will be to exclude the negative influence of people from the forest ecosystems and harness their positive energies to enhance ecosystem quality and productive potential. It will involve more proactive popular participation than the hitherto employed administrative participation alone. Working Plan Code 2014 requires the working plan officer to discuss the management issues with the people.

Forest Governance Reforms for Forest Management

A shift of management style will require the incorporation of any new approach in policies, development planning and the working-plan system. The National Forest Policy of 1988 lays stress on achieving ecological security and environmental balance. An assessment of forest governance will be necessary to understand strengths and weaknesses and also to identify changes that would be required for implementing the changed forest-management processes. A shift to an improved management system will be a long-drawn process. Therefore, the principle of adaptive management will have to be adopted and practised so that a flexible process is set in motion, and corrections and improvements may be made periodically as the learning curve rises.

Management Planning

The forest-management plans are traditionally called 'working plans' as their preparation was taken up during the colonial era of the nineteenth century with the main purpose of 'working' the forests. In its actual implementation, it was reduced to an instrument for the exploitation or mining of forest products, mainly timber. The foresters are very proud of the working-plan system, in that they assert that this was a pioneering approach in India at a time when no planning was practised in any sector of the economy.

The National Working Plan Code–2014

The National Working Plan Code–2014 (NWPC–2014) is significantly different and forward-looking compared to its precursor, the NWPC–2004, which was more of a vehicle to carry forward the forest-planning traditions from the past. The NWPC–2014 not only includes guidance on how to prepare and organize a working plan, but it also postulates the basic principles of and approaches to forest management in the country. It recognizes in its preamble that there is a paradigm shift in the perceptions about the forestry sector globally and nationally, and that the focus has shifted to the inclusion of environmental and socio-economic considerations in addition to the production of goods. 'Sustainability' is the core theme of the NWPC–2014 and the management of forests on a sustainable basis is the main goal to be realized through planning.

Objectives and strategy

The NWPC–2014 explicitly stipulates that planning for forest management must seek to manage forests and their biodiversity on a sustainable basis, incorporating ecological (environmental), economic (production) and social (including cultural) considerations. This goal is to be achieved through a number of strategic objectives that can be grouped into four categories (MOEF 2014):

1. Conservation
   (a) Conserving forests
   (b) Reducing forest degradation
2. Maintenance and enhancement of ecosystem services
   (a) Maintenance and enhancement of ecosystem services, including ecotourism

    (b) Maintenance of biological diversity
    (c) Increasing carbon sequestration potential
    (d) Improvement and regulation of hydrological regimes
    (e) Prevention of soil erosion and stabilization of the terrain

3. Production of goods
    (a) Enhancement of forest productivity, together with establishment of regeneration to improve forest health and vitality as per ecological and silvicultural requirements of the species
    (b) Progressively increasing the growing stock
    (c) Sustainable yield of forest produce

4. People
    (a) People's involvement in planning and management of forests
    (b) Fulfilling socio-economic and livelihood needs of the people.

Essential features

The following are the essential features of planning and management which are an integral part of a working plan:

    (a) Biodiversity conservation and development
    (b) Forest fires and protection
    (c) Forest health and disease control
    (d) Soil and water conservation
    (e) Water-resource management
    (f) Forests and climate change
    (g) Carbon sequestration and mitigation
    (h) REDD+
    (i) JFM
    (j) Community forest management
    (k) Fringe forest management
    (l) Application of modern technologies
    (m) Forest surveys, mapping and inventory
    (n) Grid-based sampling designing
    (o) Growth data and carbon sequestration
    (p) Linkages with forest inventory
    (q) Trees outside forests (TOF)

Improvement over previous code

The NWPC–2014 is more comprehensive than the previous code. Against the earlier working plan design that had seven chapters, NWPC–2014 suggests 11 chapters to focus on the changed emphasis of management objectives. Thus, without being totally disconnected from the past, the code ensures that the paradigm shift in the management of forests for multiple objectives and multiple uses is reflected in the working plans prepared under the NWPC–2014. The emphasis is on integrated and ecosystem-based management, with a view to ensure flow of goods and services on a sustainable basis.

Capacity building

While the NWPC–2014 is quite comprehensive, the working-plan officers, when they take up this assignment for the first time in their careers, may face difficulties with conceptual clarity and the use of scientific techniques, tools and methods for measurements, assessments and analyses of various parameters relating to growing stock, biodiversity, social impacts, and so on. This necessitates the capacity building of SFD officers through training, field visits and demonstrations.

Integration with policy

The management principles that have been prescribed in the working plan code must find a place in the national forest policy. The code reflects changes that are desired in forest management, without explaining how these changes are to be introduced. The 'paradigm shift' is only rhetoric for now, as there is inadequate understanding among forestry establishments about the concept of sustainable management of forests that is being pushed through the code.

Prospects

With economic liberalization and reforms, there has been an impressive increase in the country's gross domestic product (GDP), but the forestry sector was not integrated in the new economic growth trajectory. Stagnation and inertia set in deeply and had a cumulative effect during the last two decades. It is a policy aberration that forestry does not find its due place in the national development strategies and economic policies. For the last two decades, economic policy has focused more on industrial and financial institutions. This apathy does not allow a country to fully exploit the economic potential of its forestry sector. This sector has also become slightly dormant for the last two decades. In this situation, environmental and conservation issues are construed as hurdles to economic growth and not as pillars of sustainable development. Judicial activism has sought to save the forests from arbitrary management and exploitation. The agencies in the forestry sector have not been able to come up with an institutional framework that can ensure sustainable management and use of forests despite having the necessary capability and capacity.

This calls for a review and rethinking of the whole issue afresh. The SFDs are strong institutions, and so are the forestry research establishments. All together, these institutions can join hands and can develop improved silvicultural technologies and further upgrade them. New strategies and policies are needed to develop and manage forests to meet the present and future needs of an ever-growing population and at the same time to expand carbon sequestration and other environmental services from forests on a sustainable basis. Rather than getting into a complex and cumbersome system of criteria and indicators, a simple and straightforward approach to sustainable management of resources is not only possible but is also desirable.

Social forestry as a comprehensive strategy was adopted in the 1970s and 1980s for expanding forest cover in the country on all kind of lands, but was relegated to a secondary position with the advent of JFM. JFM was applied to public forests only. It has been an arrangement for the protection of forests by local communities in lieu of a commitment of sharing the forest produce on harvest. This programme has been partially successful but unsustainable. It was project-, fund- and SFD-driven. Once project or funds were withdrawn, local communities as well as the SFD abandoned the initiative and the village-level initiative collapsed. Undoubtedly, JFM showed success in many areas and local people reaped substantial benefits. But it lacked the appeal and support of the masses and was inherently incapable of becoming a mass movement.

If the SFDs raised 60 per cent of the total plantations and the rest were raised by other departments in rural and urban areas and by landowners and farmers on their land, since the adoption of the 1988 NFP, at least 24 million hectares would count as reforestation or afforestation of forest land. But this has also not been captured in the FSI reports. If this were a reality, open forests should have become very dense forests by now. What really happened on the ground? No one in the government is forthcoming with an answer. How taxpayers' money has been used by the government agencies should be probed to see what actually happened with the substantial financial inputs and what work was done? How great was the failure? How has monitoring been done over the years? How was accountability fixed? The central and state governments should come forward to answer these questions so that lessons could be learned and future scams could be avoided. How did the central Ministry of Environment and Forests (MOEF) set the target and what was the efficacy of the monitoring mechanism? Did it just receive reports for compilation, did it answer questions in Parliament and did it do any meaningful work?

Plantation Quality and Productivity

The SFDs are responsible for managing the forests. They undertake afforestation and reforestation in their respective states. Not only do they carry out field operations to regenerate degraded forests through sowing, planting and tending operations, they also plant trees on totally empty and barren legally defined forest lands. Plantation is carried out at the onset of the monsoons, except in some Himalayan temperate regions, where both autumn and spring planting are done. The seedlings for planting are brought from nurseries established and maintained by the SFDs. The two major lacunae in nursery production are (a) seeds are not collected from good sources and (b) the unhealthy seedlings are not culled in the nurseries but are also transplanted. Poor-quality seed and seedlings result in poor-quality saplings and trees. They may not be disease- and drought-resistant varieties either. As a result, there is generally high mortality.

The SFDs undertake casualty replacement in the same season, but the major mortalities take place when the summer comes. Either watering is done in the summer or it is inadequate or impossible due to remoteness. Not even 10 per cent of forest plantations are watered. Irrigation of plantations are a rare phenomenon in the work culture of the SFDs. In a majority of cases, there is a budget constraint and irrigation makes costs too high. The SFDs report only plantation progress, but never report mortality and survival. If the subsequent years' maintenance costs are not provided in the budget, no follow-up action is taken. In operations such as assisted natural regeneration or enrichment planting to enhance stocking, there are situations where follow-up operations are not included in the budget and are therefore not undertaken. The SFDs thus fail in proper budgeting and undertaking of required operations. Every year, there are reports that plantations have been damaged by grazing and browsing by local domestic cattle. The SFDs call it 'biotic interference', which has traditionally remained a standard excuse. The SFDs have failed miserably to address the grazing issue despite the clear policy of 1988 and supporting laws. Even if politics is involved, the SFDs are expected to come up with innovative approaches and programmes.

There is an obvious governance failure in that the SFDs lack innovations and initiatives. Their focus is on the protection of trees from illegal cutting and encroachment of forest land, and even in that, they face challenges and difficulties. Their other focus is meeting the set targets of the forestry-development programmes, which they implement but without bothering about outcomes. The accountability of the institutions remains a major issue so is the continued focus on output rather than outcome and impact.

Part III

Approaches Adopted in the Past and Their Outcomes

CHAPTER 6

Social Forestry, Afforestation and Wastelands Development

Social forestry began in the mid-1970s and grew as a movement and campaign in the early 1980s. The main objective of this programme was to plant trees outside forests to produce fuelwood, fodder, poles, small timber and other minor forest products with a view to reducing pressure on natural forests (Khan 1987). It also aimed to create employment in rural areas, reforest degraded forests, and extend, train and research to support this programme. The important activities of the programme were tree planting on the villages' common lands (wood lots), along roads, canals and railway lines (avenue or strip plantations), on institutional lands, in degraded forests and on farmland (agroforestry or farm forestry). For the first time, social forestry sought to involve local people and civil society in a forestry programme and reached out to farmers and other landowners to extend technical assistance and a supply of seedlings for undertaking agroforestry and tree plantations on their lands.

Both multilateral and bilateral international aid agencies extended financial support for the implementation of social-forestry programmes in the mid-1980s. These included the World Bank (WB), United States Agency for International Development (USAID), Swedish International Development Cooperation Agency (SIDA), Canadian International Development Agency (CIDA), Department for International Development (DFID, UK), Japan International Cooperation Agency (JICA) and the European Union (EU). In 1985, for example, 10 externally aided social-forestry projects with a value of more than US$400 million were under implementation. It became a best practice, and developing countries sought to learn from India about this programme.

In the first phase of forestry in India, the WB supported only one project, the Madhya Pradesh Forestry Technical Assistance Project approved in 1975. This project was primarily developed for plantations for the pulp and paper industry. During the social-forestry phase, the WB supported seven projects in various states (nine in all) for a total of US$345 million. The WB approved its first social-forestry project in 1979 and the last one in 1985. There was a break in the approval of new forestry projects for India between 1986 and 1992, though implementation of the social-forestry projects continued till 1992. Between 1979 and 1985, one project was approved every year.

The social forestry programme appeared to find greater acceptance among stakeholders than the subsequent joint or participatory forest management, which started in the early 1990s and reached its peak within a decade. Social forestry also found its place in India's National Forest Policy of 1988 (MOEF 1988), which called for building a peoples' movement for conservation, afforestation and social forestry. Social forestry was a buzzword in the 1980s. Substantial funds for social forestry came from donor agencies, the rural development sector as well as from national development plans, in unprecedented amounts.

The National Forest Policy discouraged industries from depending on public forests for their raw-material needs and encouraged them to obtain these from agroforestry sources. Demand for softwood increased from agroforestry and farm forestry plantations for pulpwood and veneer-quality wood (MOEF 1988). Mainly eucalypts, poplars, casuarina and *Acacia auriculiformis* were planted on agricultural land to a large extent and were easily sold to the industry. The government's role remained that of extension and providing seedling supplies, and market forces took care of the rest, including prices.

Impact of Social Forestry

Since Independence, the highest rate of afforestation was in the period under the Seventh Five-Year Plan (1985–1990), when 8.86 million hectares were planted under the social-forestry programme at an average rate of 1.772 million hectares per year. Until 1979, the cumulative plantation since 1950 was 3.55 million hectares. In the 1980s, there was a boom in plantation establishment, with the launching of social-forestry projects in many states and as part of rural development schemes (MOEF 1999a). Various organizations and individual researchers carried out a number of evaluations and research studies. The main finding or conclusions were that social forestry produced the following positive outcomes:

(a) Successful plantations in a large area
(b) Increased production of biomass, small timber, fuelwood and fodder
(c) Increased production of industrial raw material (mainly pulpwood)
(d) Greening of barren lands and aesthetic plantations in strips (avenues) along roads, canals and railway lines
(e) Restoration of degraded forests
(f) Additional income for farmers
(g) Employment generation in rural areas
(h) Increased awareness about tree planting and environmental issues among people.

The financial and economic analyses of various projects (World Bank 1994) revealed that agroforestry provided attractive returns. For example, a post-evaluation report of the WB National Social Forestry Project (implemented in Gujarat, Himachal Pradesh, Rajasthan and Uttar Pradesh) provided the following results of financial and economic analysis (Table 6.1):

Table 6.1 *Financial and Economic Analysis of Agroforestry*

State	Farm forestry: Rate of return (%)	Community woodlots: Rate of return (%)	Rehabilitation of degraded forests: Rate return (%)	Strip plan-tations: Rate of return (%)	Total project: Rate of return (%)
Gujarat	31.3	9.2	13.98	1.7	23
Himachal Pradesh	25.7	20.8	14.9		13
Rajasthan	18.4	13.99	15.6	6.6	12.1
Uttar Pradesh	43.3	13.8	13.9	10.2	27.7
All states					22

Source: World Bank (1994).

This shows that agroforestry has been financially a lucrative activity. The rate of return on wastelands or degraded lands planted in rain-fed conditions was bound to be lower. The investment in social forestry proved to be economically advantageous for the country and certainly transformed our landscapes and wood balances (Kumar et al. 2000). But it was suddenly slowed down and its full potential was not allowed to be significantly utilized to contribute to rural poverty reduction.

In the early 1990s, external donor agencies phased out financial assistance to social forestry. The rural development sector excluded mandatory allocation for afforestation under its central schemes. The states did not provide adequate budgets to continue the programme at the same or even a slightly lower level. So the entire social-forestry programme had a near-vertical collapse in general and in states dependent on external funding in particular. The newly created social forestry wings in many states had to even struggle for staff salaries. The major achievements of the programme were as follows:

(a)   The social-forestry projects increased awareness about the importance of tree plantations outside the forests
(b)   Social forestry resulted in increasing the supply of wood products
(c)   The projects encouraged the isolated forest departments to accept that local people's cooperation could be sought in increasing tree cover

(d) They demonstrated a direction through which this country could stabilize its tree cover despite extreme pressures from huge human and livestock population

(e) Implementation of projects also brought to light many distortions and imperfections in the wood market and the legal and procedural framework that had made cutting of trees on private lands, their transport and their sale difficult and cumbersome

(f) The programme had a positive impact on water conservation, improved microclimate and helped in improving soil fertility

(g) Farm forestry brought income to farmers and intermediaries and also supplied raw material to wood-based industries

(h) Plantations on farm bunds arrested soil erosion and gully formation.

Technical Constraints and Productivity

Among the various constraints of this programme that have been pointed out, one important weakness was that the massive effort was inadequately supported by research and improved silvicultural practices. Others include:

(a) The main focus was on using whatever planting stock was available on the sites and on meeting targets, that is, on quantity and not on quality.

(b) Productivity was not given the required attention—the seedlings planted produced varying qualities of plantations.

(c) The most important gaps were the quality of planting materials, lack of appropriate models for regenerating degraded forests, planting practices, and lack of models for commercial and farm forestry production.

Planting stock and planting practices

It is easy to improve yields of plantations by using better tree seeds and better nursery-management practices. There are two aspects to the issue of quality of planting materials—the genetic constitution and the physical quality of the plants. In terms of inputs and time, upgrading nursery practices is the quickest and easiest route to improving planting materials. Genetic improvements to plant potential are mainly made through long-term tree breeding.

Improving the quality of planting material was identified as one of the most important factors to improving productivity and quality of forest plantations on both public and private lands. Therefore, social and farm forestry required improved silvicultural techniques that ensured sustainability and improved productivity. Development and use of new or improved silvicultural practices appropriate to specific sites remained the most important weakness. Selection of site, site–species matching, species mixture, inter-tree spacing, soil working, post-planting maintenance operations and so on were given due importance.

The physical aspects of seed quality could be significantly improved over the short run, which could have also ensured to some extent the genetic quality of the planting stock. Generally, tree seed collection and handling as well as nursery management have not been accorded sufficient attention (World Bank 1993). At present, seed harvesting continues along traditional lines, with field staff concentrating on collecting seeds in the appropriate seasons. Plantation planning and stocking can be improved by organizing phenological studies over the year in order to predict the time of fruit ripening accurately for individual harvesting locations. They should also be able to estimate the anticipated quantity of seed which will become available at harvest time. This approach has the added benefit of indicating any unusual pathogen activity and also the responses of different trees to climate, affecting their seed set and so on.

Improvement of planting stock was necessary to select and even identify varieties of species through seed selection from candidate plus trees, designated seed-production areas and seed orchards, vegetative propagation and use of clonal plants, and improved nursery practices focusing on better containers and potting mixture. The issues of seeds, seedlings and clonal planting materials; availability of nursery containers and potting mixture; plantation models, site–species matching, species mixture, spacing, aftercare and inputs such as watering, weeding-cum-hoeing and thinning; and more efficient plantation management did not receive due attention.

Relevance of Social Forestry

Social-forestry programmes proved to be a dynamic phenomenon, gaining momentum in a number of directions, which had not been anticipated (World Bank 1994). However, with the top-down push for joint forest management (JFM) in the early and mid-1990s, social forestry and farm forestry were relegated to a secondary position. Agroforestry or farm forestry emerged as a sustainable land-use practice. In JFM, attention was shifted to natural forest management, with community participation for forest protection and sharing of forest produce between the government and local people. However, social forestry still remains relevant as an overarching forestry development strategy.

India's 1988 policy encouraged the private sector to establish direct supply links with the farmers, yet some legal obstacles on cutting and transit regulations that are a disincentive to the private landowner remain (though some have been already removed) and need urgent attention. First, attempts need to be made to develop tree species that are suitable for planting in subsistence areas, with enormous benefits for the country. In India, where land and water resources are scarce, agroforestry and tree plantations in watersheds in rain-fed and semi-arid areas could be a part of farmers' subsistence strategies and complementary to crops. Second, by correcting for the major reasons that led to failure of the community forestry component (village wood lots), tree plantations on common lands and wastelands can be used as an effective strategy of wasteland development.

According to the *State of Forest Report 2013* (FSI 2014), the tree cover outside forests is 9.13 million hectares, compared to that in the designated forests being 69.79 million hectares, that is, 11.5 per cent of the total tree cover in India. But interestingly, 26 per cent of the growing stock is in areas outside forests. Thus, 2.8 per cent of the land produces 26 per cent of the growing stock (Table 6.2). This shows the performance and potential of productivity outside forests.

According to the National Forestry Action Programme (NFAP 1999) and the National Forestry Commission (NFC) (GOI 2006), about 25 million hectares of non-forest wastelands are available for

Table 6.2 *Growing Stock Inside and Outside State Forests*

	Million hectares	Percentage of forest and tree cover	Percentage of geographical area	Growing stock	Growing stock (million m³)	Percentage growing stock
Forest cover	69.79	88.5	21.23	Inside forests	4,173.36	73.8
Trees cover	9.13	11.5	2.78	TOF	1,484.68	26.2
Total	78.92	100	24.01		5,658.04	100

Source: FSI (2014).

tree planting, and 5 per cent of the total agricultural land (about 7 million hectares) is available for tree planting under agroforestry or farm forestry. NFAP (1999b) recommended a two-pronged strategy to expand forest cover—improve the density of forest cover density (31 million hectares) and establish plantations on non-forest lands (29 million hectares).

The NFC made a recommendation in its report (GOI 2006) that forest departments must cooperate and support agroforestry by providing quality seedlings and technical assistance. The emphasis was on tissue culture and biotechnology for producing and providing quality seeds and planting material and the provision of extension services and technological support.

The target for the Green India Mission (GIM) included planting 5 million hectares outside the designated forests (see Table 6.3).

Potential of Agroforestry

Dependence on imported wood is not sustainable in the long run as the exporting countries may reduce logging with a view to reducing deforestation to meet international climate-change mitigation obligations. The large-scale imports have only been possible as many tropical

Table 6.3 *Proposed New Forest Cover under GIM*

Land	Million hectares	Approach
Scrub, mangroves, ravines, cold desert, shifting cultivation areas, abandoned mining areas	1.8	Afforestation
Urban	0.2	Tree planting
Cultivable land	3.0	Agroforestry/ social forestry
Total	5.0	Outside forests

Source: MOEF (n.d.).

countries earn a sizeable revenue from timber exports. Secondly, India—despite its large population—still has 50 million hectares of degraded forest lands, which have the potential to produce four to five times what the net yield is today, with appropriate inputs, technology and management. Agroforestry still has great potential and, with better quality of planting material, can produce within 10 years 5–7 times what it is producing today.

Since the major motivating factor for agroforestry became commercial profit (instead of meeting the basic wood requirements of households or the local population, as visualized), through the creation of an enabling environment and research and the extension of private tree plantations outside forest areas, it can meet a large part of the demand for wood products in the country (Khan 2015).

Afforestation also increases the diversity of flora and fauna and has highly varied impacts on groundwater supplies, river flows and water quantity. Consideration would need to be given to synergies and trade-offs related to forestry activities for REDD+, in the context of broad environmental, social and economic impacts, such as (a) biodiversity; (b) quality and quantity of forest pastures, soils, and water and moisture conservation; (c) the ability to provide food, fibre, fuel and shelter; and (d) employment, human health, poverty and equity.

Incentives

Policies and programmes have not been developed to provide additional opportunities for farmers practising agroforestry by way of incentives for carbon sequestration, for short-term and long-term income generation for landowners or for sustainable use of marginal agricultural land. This is all a proven fact. What is needed is to add carbon value and incentivize the whole effort so that agroforestry can become doubly lucrative, to attract more and more landowners to tree planting. Carbon incentives can also help in encouraging long-rotation tree species, which can be in the ground for 20–30 years.

CHAPTER 7

Participatory Forest Management

Joint Forest Management: A Major Policy Instrument

In all the social-forestry projects implemented in India with financial assistance from the external donor agencies, there was an emphasis on the involvement of local communities in the programme. As far as the sharing of the benefits was concerned, the formula for the community-land plantations was that the local people were to take over the plantations after 3–5 years, and for the degraded-forest plantations, there was a sharing formula of 25 per cent of the harvest to be shared with them. An agreement with the local panchayats (village councils) was also envisaged. The elected panchayats did not exist in each state; therefore, the idea of village forest committees was promoted. However, in most project states, the arrangement remained on paper only and, barring West Bengal, it is not known what arrangements followed and what kind of benefit sharing took place. One achievement was that wherever the participatory arrangement was sincerely formalized, the protection of plantations became remarkably successful. Participatory forest management (PFM) or joint forest management (JFM) is thus another stage in the evolution of the community forestry process.

About 200 million people are engaged in forest-based activities and draw income from the collection of fuelwood, small timber, grass, bamboo, medicinal plants, fruits, fibres, gums, roots, tubers, leaves, fodder, flowers, bark, and so on. Mostly landless and poor people engage in these activities. Most of them are part-time activities undertaken by women, and they may engage in more arduous labour as well. For most of these activities, there is relatively unrestricted access.

Through community participation, people have a wider choice and develop a sense of empowerment to make better/more informed choices. By way of support activities, people get employment, revolving credit, energy, roads, schools and drinking water. The forest lands are developed or improved so that the income of poor people who have access to it would increase. Access to the market for non-timber forest products (NTFP) is improved. Many institutional impediments to market access have been removed (such as state monopoly over some NTFP).

The local village communities, panchayats and, in some areas, village forest committees exert considerable influence on forest management. Forest protection and conservation are not possible if local people continue to exploit the forests for fuelwood and fodder in an unregulated and unsustainable manner. The result of such an influence is obvious in the trend of the ongoing degradation and depletion of forest resources. In areas where the local people cooperate and participate, better conservation results are possible. The JFM arrangement that had been initiated successfully in some areas in the 1990s has now expanded to 65,000 villages. The principle of JFM is that the produce has to be shared by the local people in the successfully rehabilitated forests. Except in West Bengal, where such efforts were initiated two decades ago, sharing of harvested timber is yet to be seen, however.

Compulsions and failures to protect forests by traditional means resulted in a decision by the government to promote JFM for the regeneration of degraded forests. The local community, under this arrangement, is entitled to receive a share in the produce out of both intermediate and final harvests. The idea is that unless the local communities, having customary users' rights in the adjoining forests, cooperate in the protection of the forests from unregulated grazing and tree felling, it would be beyond the capacity of the official machinery to effectively protect, conserve, regenerate and manage forests. JFM involves many complex legal and institutional issues. Nevertheless, the JFM approach gained momentum in the 1990s and the early years of this century, and remains a major approach to forest conservation and development.

The Ministry of Environment and Forests (MOEF) gave an impetus to the participatory development and management of degraded forests by issuing a circular on 1 June 1990, which contained guidelines for the state governments. The states introduced JFM as a follow-up to the said circular. In the designs of forestry projects assisted by external donor agencies such as the World Bank (WB), Overseas Development Administration, UK (ODA), European Economic Community (EEC) and Overseas Economic Cooperation Fund (OECF, Japan), the JFM approach was adopted for degraded forest rehabilitation. The experience was drawn from successful regeneration of degraded sal forests in West Bengal, where the experiment started in 1972 (in Arabari). The MOEF issued more guidelines for JFM and financed a central scheme (the National Afforestation Programme) for promoting JFM.

JFM was introduced and expanded on the underlying principle that the benefits from the degraded forest area, developed and protected with active support and participation of the local people, would be shared with the local people. The community will also participate in the planning and management of the resource. JFM is one of the options which offers protection and development of forests, as the traditional approach has become less effective in combating deforestation. Successful JFM also involves institutional changes in the forest department, including reorientation, effective extension capability, improved technology, marketing for forest products, resolution of internal conflicts within the local communities and for inter-village conflicts, facilitating or enabling legislation, promotion of NGOs' roles and effective involvement of panchayats in the whole system.

Lessons Learnt from Externally Aided State Forestry Projects

Targeting the Poor

The forestry projects target poverty alleviation directly or indirectly through community-based forest management, another name for PFM or JFM. This finds explicit mention in the development objectives of Andhra Pradesh, Uttar Pradesh and Madhya Pradesh projects. The village communities on the forest fringes or within the forests participate in the protection and regeneration of forests adjacent to the village. All village or hamlet households have the option to participate in the programme, though the immediate benefits flow to the poorer tribal communities, landless peoples and women's groups. In the villages, forest protection committees (FPCs), eco-development committees (EDCs) or village forest committees (VFCs) were formed, with adequate representation of women in the executive or managing committee. The poor villagers who depend most on the forests for their livelihood and subsistence are the main target group of the programme. A study conducted in Andhra Pradesh revealed that the migration of villagers for employment has reduced and quality of life has improved in JFM villages, and improved forest regeneration has provided additional income from collection of NTFP.

The JFM activities also target the most disadvantaged groups living near the forests (primarily tribal communities and scheduled castes): monitoring surveys suggest that 95 per cent of the households participating in JFM are poor in absolute and relative terms. While there are differences between the regions, across the state, more than half of the members of village forestry groups are from scheduled tribes or castes (this is particularly significant as the statistics from the same areas suggest that these groups make up one third of the population).

It is difficult to assess the extent to which this successful targeting has actually led to improvements in livelihoods because of the paucity of monitoring, but case studies, anecdotal and circumstantial evidence have been encouraging.

(a) JFM investments clearly have generated considerable employment and income.
(b) There had been a significant change in migration patterns, suggesting that JFM had successfully generated sufficient income-earning opportunities to discourage seasonal migration. Researchers concluded that the JFM programme is disproportionately benefiting the poor because 'low income households with low opportunity costs are gaining additional benefits from forests due to their ability to invest more time and labour in forest exploitation'.
(c) JFM has clearly improved the living conditions of the poorest in areas where forests were previously highly degraded. The regenerating forests have produced greater quantities of accessible fuelwood and NTFPs for home consumption. In some places, the regenerating forests have resulted in significant improvements in water supplies (both for drinking water and farming), reducing the burden—particularly on women—of water collection. In other places, farmers are now able to introduce new crops or extend the growing season.

Significant asset transfer to the poor

Government orders in each of the states stipulated the benefit-sharing arrangements for participating communities, which represents a huge transfer of assets to the communities. Calculations for Madhya Pradesh alone, where about 25 per cent of forest land is now under JFM arrangements, shows that the community share of this transfer amounts to almost US$1.5 billion.

Improved Participation of Women

While each project identified the need to work with women, who are the primary users of forests, many forest departments have found this a difficult task. All the state policies recognize the importance of women, ensuring an equal right to membership in the village's forest management group and insisting on a place for women in the executive committees. But in reality, these provisions have done little to bring women to the forefront of JFM. As foresters increasingly appreciate the value of active women's participation, they are developing various approaches to tackle this problem. In Andhra Pradesh, the forest department hired a number of local village women in each district to help the foresters in promoting greater women's participation: currently they are working with 650 groups. All the states sought NGO support to improve women's participation: in Uttar Pradesh, each JFM planning team includes two women from NGOs; in Tamil Nadu, under the WB Forestry Research, Education and Extension Project (FREEP), eco-development planning was facilitated, with a number of women from the NGOs as part of the team.

In some villages, earnings from forest management contributed to the self-help group (SHG) kitty; this is then, in turn, managed by the women as a source of credit for members. In both Andhra Pradesh and Madhya Pradesh, foresters found that the establishment of SHGs had led to appreciable improvement in women's participation in the affairs of the village forest-management committees.

Weightage Accorded to Social Mobilization

The objectives of all the state-forestry projects has been to regenerate degraded forests, improve productivity and conserve biodiversity through community participation and, as such, social mobilization has been an integral and necessary element of the programme. As the bulk of investment was targeted to community-based forest management, or JFM, social mobilization and the formation of effective village-level institutions or community-based organization remain the starting point. In addition to development and conservation of forests, a small investment is made in providing amenities to the participating community, such as borewells for drinking water, small water-harvesting structures, biomass-efficient cooking stoves and biogas plants, small buildings for community centres, vocational training and so on, depending on the priorities of the people. These were termed 'entry-point activities'.

The projects focus on a single sector, though there are spin-offs into the livestock, horticulture and water-resources sectors through fodder production, pasture and silvopasture development, plantation of fruit trees and catchment protection, resulting in groundwater recharging and reduced siltation downstream. In the WB's West Bengal Forestry Project, fodder research, fodder production and livestock development programmes were also included as small components implemented by the animal husbandry department. In Maharashtra and Andhra Pradesh, fodder production was encouraged to reduce grazing pressure on forests and to promote stall-feeding. The forestry projects also included an element of horticulture, by planting fruit-yielding tree species and improvement by grafting of some fruit species such as tamarind and *aonla* (Indian gooseberry). Some village FPCs have even started a school by employing a local teacher and using the community building as the schoolhouse, and also taken up adult education programmes.

As the forest-fringe villages are in the remotest areas, most government agencies do not have much of a reach there, and most people remain deprived of education and health facilities, drinking water and roads. Forest departments, due to their presence in these areas, are able to work with people and organize delivery systems. The funding for village amenities through JFM remains inadequate, though the villages' revolving funds or development funds were augmented in some areas by sale of grass or other NTFP. Where substantial incomes start flowing—for example, from sale of bamboo—the community has access to additional funds for improving amenities in the village. The experience with JFM has been that once the community is helped to organize themselves into a VFC or FPC and empowered, they develop confidence from sharing control over their natural resources and find it easy to leverage funds and services from other sectors.

Institutional Reorientation

Change in the attitudes of government-agency staff at state, district and village forest block and beat levels has been a prerequisite to the success of JFM. This transformation took time in an institution which traditionally, for more than a century, had operated more like a police force. In the forestry projects, this process started with training and workshops to reorient the staff to the participatory approach and methods. The forest departments historically have not had the skills to work with the communities, and often had an adversarial relationship. It took some time to reorient, adjust and change the outlooks of the forestry staff to acquire skills and enable them to win the confidence of the people.

Whereas the state-level public institutions take overall responsibility for policy and planning, monitoring, allocating funds and training of staff, the programme is implemented at the district level and forest-beat (village) level by the forest department staff. Unlike many other departments, the state forest departments (SFDs) are present down to the village level, are in direct contact with the community members, and facilitate the formation of VFCs and jointly work with them. One field-level member of the forest staff participates in the VFC's executive or managing committee, helping to maintain records of proceedings, minutes of meetings and accounts and with planning and executing forest-regeneration and village-development works. There has been no need to create a separate institutional arrangement for JFM as the existing public institutions for forestry have a statewide organization, though JFM has put a tremendous amount of pressure on SFDs and is overstretching their capacity.

The panchayat raj institutions are not found suitable for JFM because of the inherent power politics and their control by rich and influential villagers, resulting in inevitable corruption, feuds and conflict situations. Panchayats cover more than one village and, in many such situations, where scheduled caste and tribal populations are in the minority, their hamlets or villages are neglected. At the same time, the panchayats cannot be ignored. Thus VFCs are village- or hamlet-based institutions, which in many areas seek patronage from the panchayats to avoid conflict, through including in the managing committee at least one office-bearer from the panchayat.

VFCs are governed by either a government order, a notification, a resolution or the Indian Forest Act or Panchayat Act. For example, in Uttar Pradesh, the VFCs are subcommittees of the panchayat and JFM is governed by the Forest Act. The experience is that informal community-based organizations (CBOs) are not effective and do not guarantee community rights over natural resources, which are legally public owned. The forest department staff also find representation on FPCs and VFCs, and in Andhra Pradesh, even NGOs are represented to make the whole arrangement transparent. These institutions as such are formal in nature and have so far remained insulated from power politics. That does not mean that conflicts do not arise, particularly when transfer of significant resources (assets) is effected in monetary terms. Therefore, it has been useful to have the forest department as a facilitator and also as a regulatory authority.

The VFCs in many areas are able to articulate the village's needs and priorities and to represent the village to get assistance and resources from other sectors and government agencies. They manage—jointly with the forest department—the natural resources, protect these, improve their productivity and distribute the benefits among beneficiary members. They also set up and manage village development funds and revolving funds. In Andhra Pradesh, women's thrift groups are also encouraged, particularly in villages dominated by tribal populations. The decision-making process is by and large through discussions and consensus.

Cost Sharing

The natural resources are publicly owned, and for regeneration of degraded forests and management of these resources, investment comes from public funds. The poorer, cash-strapped villagers are unable to contribute to costs in the form of cash. However, their contribution may be in the form of voluntary labour for the protection of the forests, by restricting access to livestock, night patrolling, nabbing organized timber smugglers and organizing.

In some villages, individual beneficiaries contribute in cash a part of the cost of an item given for income generation—for example, a sewing machine, rope-making machine, iron plough, leaves and machines for making plates or materials for basketry. This amount invariably goes to the village's revolving fund. This is an approach adopted in the villages in which eco-development program is implemented, where the focus is on income-generation activities and not on usufruct sharing.

In the forestry projects, individual beneficiary schemes are very small in proportion and restricted to areas where harvesting and consequent sharing of forest produce from protected areas (PAs) is not permissible due to the Wildlife (Protection) Act.

Capacity Building

The forestry projects support capacity building of the forest-department staff, members of village committees (VFCs, FPCs, EDCs) and NGOs through regular training programmes, workshops, study tours and group interactions. The maximum number of grassroots-level NGOs (about two hundred) are involved in Andhra Pradesh. Once CBOs develop capacity to manage forests, there may be greater devolution of control over natural resources and a much reduced role for the forest department.

So far, the organized private sector has not played any appreciable role. However, the farm forestry or agroforestry sub-sector, under which the farmers plant trees on their land, is in the private sector. Agroforestry is mainly based on short-rotation pulpwood species. The role of the forest department there is that of an extension agency, providing technology and technical advice. In the past, the forest department had also provided subsidized seedlings to small and marginal farmers. In West Bengal, group farm forestry has been very successful and became a source of income from marginal agricultural land. However, the private sector's role would inevitably be enlarged when the marketing of forest products harvested from JFM areas begins. The forest-based industries are mainly in the private sector and do not have control over raw-material supplies, which have been shrinking in the recent past when they came from the public forests, due to restrictions on commercial harvesting and removal of logging and price concessions.

Outcome

JFM has been under implementation for the last three decades. The Government of India's (GOI) Ministry of Environment, Forests and

Climate Change (MOEFCC) claimed that in 2010, 24.65 million hectares of public forests were covered by the JFM approach in 28 states, where 112,816 VFCs had been set up (MOEF 2010). The report further claims that in Tamil Nadu alone, there was an increase of 39–60 per cent in the tree density in the forests after the introduction of JFM from 1997 to 2000. However, the report is unclear about the baseline year and the expenditure incurred on regeneration and plantation in the area sampled. Therefore, the change in tree density may or may not be wholly attributed to JFM.

The JFM approach could not be implemented in areas where funds were not available as the programme has been highly dependent on budget and not community- or government-led voluntary activities. Even in the areas where JFM was successfully initiated, when the funds dried up, so did the programme. There have been successes as well as failures in JFM. The programme's partial success was because of a lack of or inadequacy of inter- and intra- mechanisms for conflict resolution, weak institutional arrangements, inadequate community participation and ownership, lack of adequate accountability and transparency, and iniquitous collaboration between SFDs and communities (Bhattacharya, Lolota and Ganesh 2009).

JFM began well, but the approach got diluted, with its expansion involving a target-driven process. The initial voluntary forest protection by the village communities was gradually replaced by paid wage labour. When projects which were providing financial support ended, the VFCs involved gave up the collective protection of forests earmarked for JFM and the forest use returned to business as usual. JFM was not sustainable. It remained a forest department-driven programme, and the responsibility for and authority over forest management was never fairly shared with the participating communities, with the latter never being an equal party. The term 'JFM' thus became a misnomer and is misleading. With post-project and post-funding inactivity, the forest-department staff discontinued visits and interactions with the local communities and their institutions set up for JFM, and resumed their old functions, going back to business as usual.

JFM has remained an institutionally fragile and inadequate intervention, as it was implemented by deeply entrenched institutions in colonial and feudal traditions, designed to achieve contrary goals (Sarin 1999, p. 55). The SFDs were not fully convinced of, nor committed to, the participation of local communities. Their traditional approach was to enforce the law against illegal activities by the local people, the activities that were offences under the forest laws. They found it inconsistent to persuade these people to cooperate in enforcing laws and voluntarily keep out of forests. These adversarial situations and relationships have been the major impediment to bringing about a desirable change that would curtail the authority of the forest departments and empower the communities. The SFDs were also not convinced that democratic decentralization of natural resource use would be a desirable approach, fearing it would discourage forest protection and conservation efforts. Sarin (1999) asserted that the JFM arrangement used the local communities as instruments for achieving the SFDs' interpretation of the forest policy objectives instead of encouraging a healthy shareholding. The GOI (2007) claimed that the concept of JFM, though initially adopted with scepticism, had taken firm root throughout the country. It further recognized that a number of issues and areas still needed attention. These included the legal status of village forestry institutions, sustainability, capacity building, convergence of development programmes, technology development and conflict resolution.

JFM has passed through many stages and yet has not reached a conclusive conceptual clarity. There was, to begin with, opposition and resistance from traditional managers and staff, and even now, a good proportion are sceptical, fearing the erosion of their authority as well as being concerned for the security and well-being of the forests and forest-land tenures. They do not believe that if left to themselves as forest custodians, the communities have the resources or the capabilities and capacities to manage such an important natural resource. This remains a valid issue. This is one reason why the control of JFM has to be retained by the forest departments.

Historically, NGOs attached a great degree of romanticism to JFM and claimed that the communities have all the wisdom, organization, capabilities and commitment to conserve the forests. They also presented JFM as a panacea. During the pinnacle of JFM activity, they became active in consultancy, advocacy, research, workshops and seminars, mostly funded by external donor agencies. That enthusiasm and activism has almost vanished and the SFDs have taken full command of JFM. The JFM policy is likely to be further weakened by the Panchayats (Extension to Scheduled Areas) Act (PESA), 1996 and the Forest Rights Act 2006.

Community Forest Management

In this approach, people are placed at the centre of development concerns at the community level, and they are fully involved and supported in achieving their own livelihood goals. To achieve this, policy and institutional arrangements would have to be changed to promote the agenda of the poor. Community forest management (CFM) would aim to improve the access of the local people to the natural resources under state control. Through CFM, tenure over assets is ensured. The participatory community is given usufruct from the forests brought under joint management and developed through investments and community protection. Almost all intermediate benefits flowing from the land, such as fuelwood, grass and fodder, and other NTFP, would belong to the community members. The community would decide to keep the whole or a part of the products for the domestic use of its members or sell the surplus. Cash inflow from the sales could be kept in a common fund to generate more employment or to distribute among members. In many villages, with initial investment for management coming from the government, the villagers have generated funds from the sale of harvested forest products, which they keep to generate more employment in the future, and thus a revolving fund starts working. Thus forest resources yield funds initially, and when bamboo, poles or other NTFP become marketable, more income flows in for gainful employment. The sustainability of livelihood thus can be attained.

JFM never evolved to this idealistic state. The WB staff who came to India with experience from Africa and Latin America, where land tenure still remains uncertain, advocated an extreme view that the SFDs should totally withdraw and hand over the forests to local communities for management. However, the SFDs would provide extension services. This view also advocated that the communities (tribal communities, in particular) should be given full power to clear forests for agricultural land use that was better for their livelihood. The inclusion of community forest rights in the FRA to some extent vindicates the WB views. However, effective devolution of total power and control over forests to communities still remains to be seen.

CHAPTER 8

Livelihoods from Forests

In the late 1990s, the World Bank (WB) began to seek justifications to remain engaged with and provide assistance to the forestry sector. As the WB became more and more vulnerable to global criticism of its relevance and continued existence, it made poverty reduction its priority. It adopted as its mission 'a world free of poverty'. As a result, the WB's strategy was totally reoriented to poverty reduction. All internal sectoral strategies focused on poverty reduction as the ultimate goal. The Andhra Pradesh Community Forest Management Project clearly spelled out poverty reduction as its main objective. The Japanese aid agency, the Japan Bank for International Cooperation (JBIC; now Japan International Cooperation Agency, or JICA) also had this dilemma and found poverty reduction was the main objective to continue to provide financial assistance to forestry in India. Poverty reduction was incorporated in the joint management approach. The WB India staff estimated that about 300 million people in India depended on forests for their sustenance needs and livelihoods. This number was subsequently adopted by all institutions, including the Government of India (GOI).

However, how forests can reduce the poverty of people living in and around them, some of whom are the poorest of the poor, needs to be examined. One view is that if forestry could alleviate poverty, these people would not be in a destitute condition in the first place. Forestry operations such as logging and planting generate wage employment but are not perpetual in a locality. Revenue sharing under joint forest management (JFM) approaches may also bring about some economic benefits that are only periodic and not sustainable.

Rural Poverty and Forestry

In India, more than 300 million people are estimated to live on the fringes or inside forests in rural areas (about 200,000 villages) and depend on the forests for their sustenance. This includes about 50 million tribal people, who are among the poorest of the poor and live in the remotest villages. They depend on the forests for their domestic energy needs (fuelwood, twigs and dry leaves), poles, small timber and bamboo for construction of their huts and agricultural implements, fodder for their livestock and a number of non-wood forest products (NWFP) such as medicinal plants, flowers, fruits, tubers, roots, gums, resins, grass, leaves, honey, and so on. Not only do they use these products for their own needs, but also earn their livelihoods by their collection and sale. They also find employment in forest-related activities, though with reduced logging, the employment opportunities have reduced and their dependence on non-timber forest products (NTFP) has increased.

With the increasing populations of both human beings and domestic animals (livestock), forests have been under increased pressure and depletion has continued. In the 1980s, the GOI launched a massive afforestation programme through social forestry interventions, mostly supported by the international donor community, both bilateral and multilateral. The WB, through the International Development Association (IDA), provided substantial financial assistance for social forestry programmes in nine states (Gujarat, Uttar Pradesh, Himachal Pradesh, Jammu and Kashmir, Rajasthan, West Bengal, Haryana, Karnataka and Kerala). This programme involved afforestation of common wastelands, strip lands along roads, canals and railway lines, and agroforestry on agricultural lands. This period witnessed a boom in farm forestry, which produced substantial quantities of pulpwood and fuelwood. It was expected that social forestry would reduce pressure on natural forests. The bilateral donors which provided assistance for social forestry, though on a small scale, included the United States Agency for International Development (USAID), the Swedish International Development Cooperation Agency (SIDA), the Department for International Development of the United Kingdom (DFID, UK) and the Canadian International Development Agency (CIDA). The Operation Evaluation Department of the World Bank (1997) audit of the National Social Forestry Project covering four states (Uttar Pradesh, Himachal Pradesh, Rajasthan and Gujarat) rated it 'satisfactory' and thought its sustainability 'likely'.

Forest-based Livelihoods

Forest-based direct and indirect employment continues to make a significant contribution to income for a large number of people, both in rural and urban areas. Forest resource management and development programmes help the poor in three ways: (a) by providing wage employment; (b) by helping them upgrade their skills; and (c) trees as assets providing subsistence needs, as a source of income and a savings bank (World Bank 2000). It is recognized today that forests can contribute to poverty reduction among the poorest of the poor villagers living in and around forest areas. However, poverty not only encompasses material deprivation, measured in terms of income or consumption, but also low levels of education and health, vulnerability and exposure to risk, voicelessness and powerlessness (World Bank 2000, 194). The sustainable livelihood framework identifies five sets of capital assets which can be built up and drawn upon by people: human, natural, financial, social and physical.

Forest-dependent livelihoods have been the oldest survival strategy for humans. In the pre-historic era, woodcutters cut trees from nearby forests to sell as fuel to economically better-off villagers and residents of nearby small towns. When the forest lands and no man's lands were acquired by feudal lords and local kings, the woodcutters continued to carry fuelwood upon their heads or on draught animals to sell in the nearby market and continued earning a livelihood. People also got employment from cutting poles and trees for house construction, furniture, bullock carts, agricultural implements, and so on. The forests have provided employment and income since time immemorial. Before India's British colonial rulers embarked upon forest conservancy in the mid-nineteenth century, local rulers, big landlords and feudal lords were allowing forest exploitation as a concession to city traders on payment of a nominal lump sum as royalty. Timber was mostly hand-sawn on the spot inside forests and poles, after debarking, were transported to nearby town markets.

In the latter half of the nineteenth century, forest surveys, demarcations and settlement processes were begun in the territories directly under British administration. Similar initiatives were also undertaken in most large princely states or kingdoms. The main objective was to harvest timber on a sustained-yield basis to meet the needs of the market, mainly for railway sleepers and the shipbuilding industry. The forest working plans prescribed an annual yield of forest products or allowable annual cut for timber, presuming that logged-over forest areas would regrow on their own. A labour force was required in remote forests for cutting, limbing and debarking trees and sawing, loading and transporting logs and sawn timber. The forest departments and logging contractors brought poor people from tribal communities to do these work, thereby laying the foundation of a large number of forest villages. This provided significant employment to poor and landless people in remote and inaccessible rural areas. The forest departments took upon themselves the responsibility of being the benevolent employer and looking after their well-being. Logging-based employment in the primary sector and in wood-based industries as the secondary sector continued for decades and still continues. Substantial employment was also generated by survey and demarcation of forest boundaries, inventories of forest resources, cleaning and thinning, construction of forest roads and logging paths, timber floating and transport, bamboo harvest, forest-land grazing, and so on. The *taungya* practice of reforestation was also introduced in many regions of India, which provided food security and income to landless families for three years by allowing reforestation with inter-space agriculture on forest lands. Plantation forestry came much later and provided additional employment. This involved collecting, cleaning, grading and treating tree seeds, preparing land for nurseries, growing tree seedlings and planting them in the fields. Big forest- or wood-based industries, including saw milling, veneers and plywood, pulp and paper, particle board, wood panelling, railway coaches, shipbuilding, furniture, resin, dyes, and so on, were also set up, creating a vast potential for employment.

Poor people derive livelihood benefits from forests in two ways: subsistence and income (see Table 8.1).

Subsistence

Local villagers collect a large number of wood and non-wood products for their domestic consumption, free of cost, and this huge extraction of forest products is not accounted for in government revenue and even in the national annual gross domestic product (GDP). The range of products varies from roots, tubers, leaves and fruits to grasses, bamboo, fibre, fuelwood, poles and timber.

Employment and income

In India, a huge amount of employment was generated in the 1980s through afforestation and social-forestry programmes funded by the

Table 8.1 *Livelihood Benefits Derived by Poor People*

Subsistence	Free collection of products	Fuel, fodder, NTFP, grazing, fruits, flowers, nuts, roots, tubers, thatching material, poles, bamboos, green manure, etc.
Income	• Employment • Collection and sale • Household enterprises	Logging, planting, tending, sawing, etc. NTFP, honey, mushrooms, fruits, gums, leaves, grass, fuelwood, etc. Making baskets, rope, leaf plates, brooms, etc.

government and a number of external donor agencies. The estimate is that 1.5 billion of person days were generated between 1980 and 1995. The land covered by afforestation and social forestry from 1980 to 1998 was about 25 million hectares, which included about 50 per cent plantations, mainly on agricultural land (MOEF 1999a). The number of seedlings distributed to farmers for tree planting on agricultural lands was about 1 billion.

The social-forestry phase gradually gave way to forest development through the JFM approach. This was a response to the fast depletion and degradation of forest resources and the inability of government agencies to effectively protect the forests, which have been becoming more and more vulnerable to illegal logging and heavy grazing, retarding natural regeneration. JFM sought to reverse this alarming trend. In the most simplistic sense, JFM is a reciprocal arrangement between the government and local communities to protect and restore forest cover on degraded forest lands and to share the resultant products and incomes. This approach was successfully tested in West Bengal and was introduced in the 1990s in Andhra Pradesh, Madhya Pradesh and other states of India, and gradually spread like a wildfire across the country. This approach is the main forestry strategy of the GOI today.

NTFP: Potential for sustainable livelihoods

During the last few years, the view that NTFP, even more than wood, can provide a considerable income to local communities on a sustainable basis is gaining momentum. As a result, the focus of policymakers and other actors in the forestry sector has gradually shifted to NTFP extraction, value addition and marketing. One of the biggest weaknesses of Indian forestry is that NTFP traditionally has not been a big source of revenue to the government, with a few exceptions such as *chir* pine resin, bidi leaves, sal seeds and some medicinal plant products that have been revenue-yielding, marketable products. The public forestry institutions were expected to have expertise in mainly forest management for timber production and extraction. NTFP was at best left free of state control and local people were allowed to collect the products to meet their subsistence needs and earn meager incomes at the village's weekly market by selling the collected products. Later, a number of states set up public-sector units to buy NTFP from primary collectors and market it with the sole objective to eliminate the middlemen or traders who tended to exploit poor gatherers by purchasing the product at a very low price. The state forest departments (SFDs) in India do not have a reliable inventory of NTFP-yielding resources in terms of extent and growing stock, and their incomplete records show only partial output figures of some NTFPs, based on the permits they issue. The SFDs also lack adequate knowledge of species of shrubs, herbs and grasses that yield NTFPs, as well as of their maintenance and propagation. They also lack information on the nature of harvesting of these products and its impact on the overall stocking and health of the forest ecosystem. The NTFP harvest has traditionally been overlooked, and it was termed as 'minor forest products' (MFP) deserving only minor interest.

There is also a lack of understanding of the impact of continuous and excessive harvesting of some NTFPs and the prevailing assumption has been that collection of NTFP can go on indefinitely on a sustained yield basis. There is also no information as to which species are fast becoming depleted due to repeated harvesting and which species have poor regenerative potential. For those species whose roots have commercial value, the whole plant is uprooted and destroyed and it never regenerates, nor are efforts made to regenerate it. The SFDs also lack an effective monitoring system to track down the impact of destructive harvesting on the resource base and on the ecosystem as a whole, which has been further compounded by a lack of baseline information. Nevertheless, it is clear that market demand is selective, which works against the ecological objective of conserving the biodiversity present in the natural forests (Arnold and Perez 1998).

Forest land for agriculture

A theory has gained ground recently that by transferring the decision on land use of forest land to local communities, poverty can be reduced and livelihoods can be enhanced. Another related view is that forest land cleared for agriculture and distributed among the rural poor would reduce poverty and hunger. More than 1 million hectares of forest land have already been illegally converted to agricultural use through encroachment, even after officially converting 2.6 million hectares of forests to agriculture uses between 1950 and 1980, and it has the potential to further expand with the periodic regularization and award of titles to encroachers. Forest land has been encroached upon by a variety of people, including powerful politicians and their protégés, people already owning land, and the landless and other poor persons. This issue is serious and has not been adequately addressed at the policy level. It entails considerable controversy and also exposes how powerful people have encroached upon, promoted and even traded in encroached lands.

The conversion of forest land for agricultural uses has necessarily been a part of economic evolution over the centuries. The negative aspect of encroachment is that it is conversion of forests in a haphazard manner, with no proper land evaluation. This leads in most cases to the creation of marginal agricultural land, with poor productivity and having the potential for remaining fallow and degenerating into wasteland after the erosion of topsoil. Creating a large number of poverty-stricken farmers is not a sound approach. At the same time, encroachment on forest lands creates inequalities and this de facto conversion of forest land to agricultural land is difficult to reverse and is politically undesirable, though there have always been efforts and pressures to legitimize it by awarding title deeds for such lands to encroachers. How much forest clearance is required for the distribution of land among the rural landless living in the vicinity of forests and dependent upon the encroached land for their livelihoods? The landless poor who have not encroached and those unfortunate counterparts who live further away from forest boundaries and lack opportunity to encroach have been and will be the immediate losers, as they would not become landowners. This goes against the very principle of social equity. The land tenure issue is a very serious one and needs to be addressed on priority, in a holistic manner and in accordance with a well-developed land-use policy.

Participatory forest management for livelihoods

One strong view is that participatory forest management (PFM) regulated by a state agency is the best way forward, as the rights and obligations of the interested parties are well defined and it is one of the most potent tools to reduce the poverty of the landless, tribal and other disadvantaged rural peoples. There is another view (held by some radical NGOs, donor agencies and academics) that state agencies should withdraw, allow the local communities to take full charge of adjacent natural resources, and authorize them to take decisions on how to manage, conserve and exploit these resources and to share among themselves the benefits or revenues. Some of the related issues which are raised from time to time are:

(a) How much forest management responsibility and control can and should be decentralized to local communities
(b) What would be the implications of such a transfer of rights, including handing over of tenure from state to local communities
(c) What would be the role of the state or public forestry institutions in the changing scenario
(d) Who are the stakeholders and who are not at the local, regional, national and global levels
(e) What would be the risk of extractive versus sustainable resource management by the local communities after the devolution of management control
(f) Conversion of forest land to agricultural land, de facto and de jure—when and where it would stop
(g) How to privatize a common public resource
(h) How to market the forest products
(i) How to regulate natural resource management

As the ongoing JFM programme has its own merits and positive impacts and is also associated with potential socio-economic conflicts, it is difficult to guess its future evolution, impacts and culmination. The emerging question is what would be the consequences of increasing emphasis on the newly rediscovered and rechristened livelihood support or poverty alleviation from forestry and how it is different from the traditionally accepted economic role and contribution of the forestry sector. Critics also question whether forestry is a good option for poverty reduction or whether there are better options. It is also suggested by many that the poverty issue could best be addressed by focusing on overall economic growth rather than pushing rural communities back to forest-dependent, 'primitive' livelihoods and thereby depriving a significant portion of the population of the benefits of the modern industrial and market economy. Poverty reduction is a fiercely debated topic in recent times, and even more difficult and controversial is poverty monitoring.

Identifying forest users and stakeholders

Forests constitute multiple-use and multiple-stakeholder resources, and one of the most complex natural systems, eluding the right formula to create a balance and often involving competing demands. First, often people who are not drawing their daily subsistence from the forests constitute a vast majority of the stakeholders. Ignoring them would have devastating consequences, because the forests release oxygen, help in maintaining atmospheric carbon balance, prevent desertification, protect important water catchments and help in maintaining local, regional and global environmental quality and stability. Second, it would be highly dangerous to fail to take a holistic view of forestry and to prefer a narrow view for the sake of convenience to achieve a short-term objective. Because the world is not yet facing an irreversible crisis today, very few people realize that forests are critical to the survival of human civilization on this earth and constitute a most critical life-support system, much like air and water. This necessitates that attempts at short-term economic, social and political gains should be made subservient to the long-term and critical needs of forest conservation.

This does not mean that forests, constituting a renewable natural resource, should not be utilized for the benefit and welfare of humankind. However, it implies that neither the local communities nor the state should have absolute power to make such drastic and catastrophic changes in forest-land use and tenure that in long run may put the very survival of humankind in jeopardy. The current state of the forests has evolved over two centuries 'from forest as no man's land to every body's land' and the forest is recognized as one of the most important public goods.

Incentives for local people to conserve forests

The legislation and regulatory framework for natural resources aims at restricting primarily local people from indulging in unregulated, unsustainable and age-old extractive habits and practices that may push the forests into irretrievable decline. The law also aims to protect forest resources from illegal commercial logging, extraction of certain forest products, fire and poaching of wild animals. Deforestation and degradation have progressed rapidly during the last five decades, with a combined effect of increased local consumption and commercial extraction of wood for state revenues. The exclusion of the impact of such activities became necessary for conservation of wildlife in legally designated wildlife sanctuaries and national parks. In the protected areas, JFM cannot be practised because of restrictions on harvesting of products or on their revenues. In many PAs and ecotourism areas, local communities have been empowered to collect entry fees and provide services to visitors. The park management also implements eco-development schemes to assist people with alternate means of earning their livelihoods, to reduce dependence on forest PAs. There is vast unexplored potential for livelihood support for local people from biodiversity conservation, and NTFP collection and value addition is one major area. The possibility of culling excess populations of wild animals is being slowly recognized, though there is strong opposition to this idea for fear of its misuse. However, it would be a far-fetched expectation to restore pre-conservancy hunting practices in forests in general and PAs in particular. The clock cannot be put back. With the implementation of conservation policies and law, that practice cannot continue and the local communities have reconciled to the changes that have been enforced during the last five decades. However, conservation efforts have resulted in serious human–animal conflicts involving loss of agricultural crops, livestock and human lives, which directly impinge upon rural livelihoods.

Livelihoods and JFM

Can JFM continue to provide sustainable livelihoods, partial or complete, to participating local communities in the long run? Will JFM be sustainable once investments in forest restoration are discontinued? Would village forestry institutions continue to thrive and sustain themselves after the investment phase and the end of wage employment? Would JFM investments yield adequate livelihood benefits to the participating community in perpetuity, so as to keep their interest alive and keep them active in protecting the forests? What has been the experience during the last three decades? (This could indicate a clear trend.) These are some of the relevant issues or questions that deserve to be analysed in depth and answered.

There are three categories of people who attempt to answer these questions in their own way. One category comprises sceptics who suspect the intentions of the government. The second includes the promoters who are engaged in implementing the programme, without much thought to its economics and sustainability. The third are the participating local communities, which are more interested in day-to-day wage earnings and income from employment in forest development. For the promoters of the programme, it is not aimed at livelihoods for the poor but to protect and regenerate vast tracts of degraded forest lands with the willing support of gradually empowered community institutions. For them, livelihoods remain a byproduct of the programme and not the main goal. Contrary to this, many influential external funding agencies are emphasizing that livelihoods should be the main outcome of JFM. It could be optimistic to assume that a change in forest-management practices would be able to provide benefits in the short term to the community. However, they also believe that with economic growth, the dependence of the local people on the forests for their livelihoods will reduce, as has been the case with more developed countries.

Impact of JFM on rural poverty reduction

This has been difficult to assess. There is no denying the fact that JFM has created vast employment opportunities and additional income for participating villages, thereby improving the quality of life therein. Not only this, but with the availability of employment next door, migration from villages has reduced considerably, improving quality of life and enabling families to send children to school. Through JFM, it is claimed, the local communities have been empowered and have developed a sense of ownership and a stake in the jointly managed forests. This impact was observed in the Andhra Pradesh (and Telangana now) state of India and has been confirmed through a number of independent studies conducted by local universities and NGOs. Investment in regenerating and restoring the productivity of degraded forests has had an enormous positive impact on rural income generation in the short term, thereby immediately improving quality of life, empowering communities and building substantial social capital. The multiplier and acceleration effects have also been observed. In many villages, communities earned a cash income by selling surplus fuelwood or fodder. This cash income was not distributed but was used to create more employment in adjoining forest plots. Thus villagers got sustainable employment for more than one year, with the first year's investment coming from state sources. Increased production of fruits such as custard apple and Indian gooseberry (*aonla*) was also recorded in Andhra Pradesh.

Poverty Reduction through Forestry

How much poverty reduction can be achieved by forestry? This remains an elusive question. There is dearth of case studies and quantifiable data and analyses to reveal the actual contribution of forestry activities to reducing rural poverty. Most research and studies on this subject have been carried out in Africa and many people tend to extend those findings to India. India is a large country and the situation varies from one region to the other and across various localities within one region. It is also noteworthy that India's record in forest conservation is remarkable considering the vast population, scarce land resources and formidable pressures on land and natural resources. Dense forests and PAs may provide some income by part time and seasonal collection of NTFP if there is a demand or market for one or more such products. However, the fact remains that unless forest resources are harvested and regenerated, there would be insignificant or negligible flow of income from the forests to the poor local people. If forests could reduce poverty, the forest dwellers and forest-fringe dwellers would not be the poorest of the poor, living in indescribable and pathetic situations. Unemployment, lack of year-round employment and the cultivation of small, rain-fed farm holdings are some of the main causes of poverty. Poverty is here defined as the lack of economic opportunity—opportunities to which people living in remote areas, particularly those living in the vicinity of forests, have no access. Wunder (2001) argues that there are few win–win synergies between forests and poverty reduction, as they may lack comparative advantages for poverty alleviation. He further contends that natural forests serve as 'safety nets' for the rural poor, but it proves difficult to raise producer benefits significantly. Urban consumer benefits from the forests are limited and generally do not favour the poor.

The WB (2001) recognized that strategies addressing poverty reduction through forests can generate internal conflicts if not implemented correctly, due to increased competition for essential forest products which may deny access to the poorest of the poor. The potential for forest-led development to alleviate a country's poverty is likely to be limited. Existing benefits to poor rural communities from forests may be protected as a 'defensive' strategy, but it is difficult to raise the benefits in a sustained manner (Wunder 2001). It would be too ambitious and unrealistic to assume that forestry alone could be a potent means of assuring sustainable rural livelihoods. The forests would always be looked upon as a source of multiple-use products and critical environmental services and would have to be maintained to serve as a life-support system for the planet Earth.

Part IV

Forest Legislation and Governance

CHAPTER 9

Legislation as an Instrument of Forest Policy

The Indian Forest Act, 1927

The overarching framework for forest management in India is governed by the Indian Forest Act of 1927 and its variants in the states. It primarily seeks to consolidate the laws relating to offences, penal provisions, powers, transit of forest produce and the duty leviable on timber and other forest produce. Historically, the trend has been towards expanding the total area of the reserved forests and reducing the area of protected forests and the so-called 'unclassed' forests. This has been legalized through the principle of eminent domain. The right of eminent domain is the power of the state to take private property for public use with the payment of compensation. The wastelands in India from which forests were carved out were treated as 'no man's land' and, as such, their acquisition by the colonial regime did not involve any payment; only through the process of land settlements under the revenue system were the rights of the people recorded.

One of the shortcomings of the Indian Forest Act is that it does not include a definition of 'forest' or 'forest land'. The Supreme Court defined the term 'forest' as per the dictionary meaning. It led to a flurry of interventions in the court due to its wide scope. 'Forest', according to the said definition, includes land which might be private, common pastureland or cultivable land. The Indian Forest Act establishes three classes of forests. The most restricted class is 'reserved forests'. In reserved forests, most uses are prohibited unless specifically allowed by a forest settlement officer in the course of 'settlement'. In 'protected forests', the government retains the power to make rules regarding the use of the forests, but in the absence of such rules, most practices are allowed. Among others, the state retains the power to reserve specific tree species in protected forests, which has been used to establish state control over trees whose timber, fruit or other non-wood products have revenue-raising potential. A third category is 'village forests', in which a state government may assign to 'any village-community the rights of Government to or over any land which has been constituted a reserved forest'.

The Act confers wide legal powers upon the administrative authorities, both forest and police departments, to take cognizance of offences relating to forests and for the grant or refusal of permits relating to transit of timber and other forest produce. This Act and its state variants have been instrumental in protecting forests from activities prohibited by law, such as illegal cutting of trees, encroachment upon forest land, grazing, kindling fires and unauthorized transport of forest produce. This has been a very effective law that has helped save millions of hectares of state forests. However, prosecution has often not resulted in the conviction of the accused. Prosecution is difficult as the state forest departments (SFDs) do not have separate prosecutors in the districts, and in the judicial process, a lenient view is taken. Nevertheless the law has been a deterrent, enabling the SFDs to discharge their responsibilities.

Legal Status of Working Plans and Micro-plans

Though prepared in accordance with the principles and methodology contained in the national code for working plans, a working plan is a forest-management guiding document for a forest division for a period of 10 years. This includes detailed information on the forest division, such as the status of forests, past management, statistics, forest produce, staff and labour supply, wildlife management, growing stock, allowable annual harvest of forest produce and other details. A working plan is prepared by the SFD and approved by the central government.

The working plans or micro-plans are not legal documents but only guiding documents to facilitate the management of forests. However, the working-plan guidelines—or 'prescriptions', as they are termed—relating to the allowable annual cut of timber prevents excessive exploitation, though permissible deviations are allowed. However, working plans as a regulatory system have historically not provided adequate safeguards against the unsustainable use of forest resources. In the Godavarman case (T. N. Godavarman Thirumulpad vs Union of India [CWP 202 of 1996]), the Supreme Court of India emphasized the role of working plans. In one of the interim orders in 1996, the Supreme Court directed that no tree felling shall be carried out in any state except in accordance with the provisions of the working plans. This forced the SFDs to update working plans expeditiously and get approved from the regional offices of the Ministry of Environment, Forest and Climate Change (MOEFCC).

The Wildlife (Protection) Act, 1972

The Wildlife (Protection) Act, 1972, is the country's single most significant law on wildlife conservation. It provides for the protection of wild animals, birds and plants and for related matters. Under it, over 560 national parks and wildlife sanctuaries—termed 'protected areas' (PAs) in common parlance (though this is not a legal term)—have been created or given legal status. Though there were several laws relating to wildlife prior to 1972, this Act was India's first comprehensive legislation covering the whole country. The Act provides for two categories of PAs, that is, national parks and sanctuaries, which can include reserve forests, protected forests and revenue lands, including common lands and even private (mostly agricultural and plantation) land.

The Wildlife Act, since its inception, has resulted in a number of legal disputes, with far-reaching implications. There is serous criticism of this Act from time to time. The main criticism is that it restricts people's access and prevents them from exercising their traditional and customary rights and favours only the wild animals, thus creating a conflict and making an issue of the human rights of the forest dwellers. An important case relating to this was the WWF–India vs Union of India Settlement of Rights case (CWP 337 of 1995). The Supreme Court, since the institution of this case, expressed concern and issued several directions for expediting the process of the settlement of rights. However, it is relevant to note that there exist no clear guidelines for the settlement of rights. The question before the government has been the question of the rights of forest dwellers, especially those living in PAs.

The SFDs have not been able to complete the process of final notification of PAs as well as the settlement of rights and the relocation and rehabilitation of people living inside national parks. Attempts were made under the direction of the Supreme Court to complete the process. However, in the meanwhile, the Forest Rights Act (FRA), 2006, was promulgated. The FRA made PA conservation subordinate to the rights of the forest dwellers. The purposes of the FRA and the Wildlife Act thus appear to be in conflict with respect to the PAs.

The Forest (Conservation) Act, 1980

The Forest (Conservation) Act (FCA) of 1980 came as a resolution by the central government to restrict the reckless deforestation caused by the indiscriminate use of forest lands for non-forest purposes. Under the FCA (amended in 1988), no state government can permit de-reservation of reserve forests and allow use of forest lands for non-forest purposes without the prior approval of the central government. The FCA does not itself ban any non-forest activity, or even the de-reservation of reserved land, leasing of forest land or clear-cutting of natural forests. All it requires is that the prior approval of the central government be obtained for such actions. The FCA has been instrumental in reducing the rate of deforestation in India. However, the criteria for permitting diversion of forest lands to non-forest uses have not been transparent.

Rights of Indigenous People under International Law

The special relationship between indigenous peoples and the natural environment they inhabit has been recognized by the United Nations. The International Labour Organization Convention (No. 69) enjoins the international community to take responsibility for developing, with the participation of the peoples concerned, coordinated and systematic action plans to protect the rights of these peoples and to guarantee respect for their integrity. Such action includes measures for:

(a) ensuring that members of these peoples benefit on an equal footing from the rights and opportunities which national laws and regulations grant to other members of the population
(b) promoting the full realization of the social, economic and cultural rights of these peoples with respect to their social and cultural identity, their customs and traditions and their institutions
(c) assisting the members of the peoples concerned to eliminate socio-economic gaps that may exist between indigenous peoples and other members of the national community, in a manner compatible with their aspirations and ways of life.

The rights of ownership and possession that the peoples concerned enjoyed over the lands which they traditionally occupied should be recognized. In addition, measures should be taken in appropriate cases to safeguard the right of the peoples concerned to use lands not exclusively occupied by them, but to which they have traditionally had access for their subsistence and traditional activities. Particular attention should be paid to the situation of nomadic peoples and shifting cultivators in this respect. The rights of the indigenous peoples to the natural resources pertaining to their lands shall be specially safeguarded. These rights include the rights of these peoples to participate in the use, management and conservation of these resources.

Principle 22 of the Rio Declaration on Environment and Development (1992) also addressed this issue, and stated:

> Indigenous people and their communities and other local communities have a vital role in environmental management and development because of their knowledge and traditional practices. States should recognise and duly support their identity, culture and interests and enable their effective participation in the achievement of sustainable development.

On 16 May 1994, an international group of experts on human rights and environmental protection convened at the United Nations in Geneva and drafted the first-ever Draft Declaration of Principles on Human Rights and the Environment. The relevant parts of the draft state that everyone has the right to benefit equitably from conservation and sustainable use of nature and natural resources for cultural, ecological, educational, health, livelihood, recreational, spiritual or other purposes. This includes ecologically sound access to nature. Everyone has the right to the preservation of unique sites, consistent with the fundamental rights of persons or groups living in the area. Indigenous peoples have the right to control their lands, territories and natural resources and to maintain their traditional way of life. This includes the right to security in the enjoyment of their means of subsistence. Indigenous peoples have the right to protection against any action or course of conduct that may result in the destruction or degradation of their territories, including land, air, water, sea ice, wildlife or other resources.

The foregoing facts clearly show the international community's resolve to protect the rights of indigenous and tribal peoples. Article 51 of the Constitution of India states that 'the State shall endeavor to... foster respect for international law and treaty obligations in the dealing of organized peoples with one another'. The Government of India (GOI) appears to have enacted the Forest Rights Act (FRA, 2006) to fulfil its obligations under the international conventions.

Forest Law Implementation

Law enforcement in letter and spirit has been a major governance issue in the forest sector. Forest protection and conservation involve serious conflicts. The state has an obligation to manage and preserve forests for the present and future generations and to increase productivity for sustainable use. The people treat forest lands as no man's land and as a common property and resource. The communities living within and around the forests claim rights of access to and use of these forests, as they depend on the forests for their sustenance. These competing claims result in several conflicts. These conflicts include those between the forest officials and local inhabitants, those between people and animals and those between people within and outside the forest areas.

Forest managers experience complex legal problems in prosecuting their cases in courts. Often, these problems are deterrents for forest and wildlife managers. The first major problem is the growing number of arrears relating to forest and wildlife cases in the courts. This is further compounded by the increasing number of cases that are reported on a daily basis at the field level. The sheer quantum of such cases is alarming and the worsening situation needs urgent attention. The rate of conviction in forest- and wildlife-related offences is minimal. There are a number of constraints to the enforcement of forest laws that the administrative machinery experiences. These include, but are not limited to, procedural delays, lack of trained legal staff, the absence of special courts to try offences against forest laws, unrepresented or minimally presented cases and the very low priority that courts give to offences against forests and wildlife.

Human–Animal Conflict

Human–animal conflicts pose another serious concern, especially from the communities' viewpoint. This mainly relates to crop damage and the loss of cattle and human lives. The rules made under the Forest Act and the Wildlife (Protection) Act prescribe the amount of compensation due in case of any such damage to life and property. However, there is no uniform rate of compensation, a situation further aggravated by delay in disbursement of compensation. The conflict leads people to poison wild animals and cut trees illegally. The offences inside PAs do not require independent witnesses. Special provisions regarding the recording of evidence and its admissibility in a court of law have been provided. This has resulted in cases of false implication due to non-availability of witnesses.

Balancing Competing Claims

The courts have played an important role in balancing the economic and ecological concerns arising out of the management of forests. Though they have sometimes been criticized for doing this, on the grounds that by undertaking the task of reconciling conflicting claims, the courts have gone beyond their mandate and have taken over a task that essentially belongs to the policymaking authorities, that is, the executive wing of the government. However, the fact that the courts have been able to balance the two types of claims on forest resources shows that there is enough legal space provided within the existing legal framework to allow for the interpretation of the law by the courts. For example, in the State of Tripura vs Sudhir Kumar Ranjan Nath case, the apex court, while examining the objectives of the Indian Forest Act, departed from the general practice of seeing the Act as primarily a revenue-oriented legislation and considered the Act as one 'to preserve, protect and promote the forest wealth in the interest of the nation', saying that it was for this purpose that the Act vests in the government the control over forests and other areas which are not the property of the government (Upadhyay and Upadhyay 2002, p. 574).

The Forest Rights Act, 2006

The British Colonial regime appropriated most of India's forests and ruthlessly suppressed shifting cultivation and other tribal rights, resulting in the devastation of tribal lifestyles, economies and cultures, forcing large numbers into wage labour and even bonded labour (a kind of slavery) (Sarin 1999). Some regions even witnessed violent rebellions against the British Raj for usurping their lands and denying them access. Thousands of tribal people were brought in by the SFDs for logging operations in the forests. Their settlements have been known as forest villages. They were provided land for their hutments and to cultivate for their basic needs. Over the years, the condition of these villages deteriorated and they were deprived of land tenure and a means of livelihood, forcing them into poverty, hunger and disease. Because of lack of ownership over the land they possessed, they could not take advantage of government subsidies, credit for agriculture and other related benefits that accrued to other non-tribal landowners. As the population in these villages increased and more forests were cleared illegally for accommodating the increasing population, there erupted conflicts. The state agencies treated them as encroachers or illegal occupants of the forest lands who had to be dealt with through appropriate legal action. Their responses involved agitations and even violence. Many activist NGOs took up their cause and fought for their rights. There were prolonged deliberations over this issue for decades and even a formula presented in 1983 to give the forest villagers heritable but inalienable rights over land in their possession. However, as it would require de-reservation and conversion of legally defined forest lands into non-forest lands, the FCA was also an impediment to this. Thus the resolution of the issue was only deferred and has not been addressed despite volatile situations in some parts of Central India, though the government had acknowledged that the tribal peoples were not given a fair deal and that they were living in a miserable state in an economically emerging India. The tribal communities, despite numerous rights and concessions given to them by the governments under welfare programmes, remain deprived of the benefits due to apathy of the government agencies (GOI 2007a).

To address this issue, the central government promulgated the Scheduled Tribes and other Traditional Forest Dwellers (Recognition of Forest Rights) Act, 2006, what we have already referred to as the FRA, as it is popularly called the Forest Rights Act. The raison d'être for this law is that the forest rights as applied to tribal communities' ancestral lands and their habitat were not adequately recognized in the consolidation of state forests during the colonial period as well as in independent India, resulting in a historical injustice to the forest-dwelling scheduled tribes and other traditional forest dwellers, who are integral to the very survival and sustainability of forest ecosystems. It, therefore, became necessary to address the longstanding insecurity of tenure and access rights of forest-dwelling scheduled tribes and other forest dwellers, including those who were forced to relocate their dwellings due to state development interventions (GOI 2006).

The FRA superseded the provisions of all other laws in force at the time of its enactment, including the Forest Act of 1927, the Wildlife Act of 1972 and the FCA of 1980, which conflicted with the former. The FRA put a halt to the relocation and resettlement of communities from the PAs. Under the FRA, it became obligatory for the central government to divert forest lands for the creation of roads, canals, schools, shops and other facilities. The forest rights of these communities have been defined and recognized under the FRA 2006. These rights include land possession, use and ownership of NTFP, grazing, fishing and conversion of pattas (or leases, or grants) issued to them into titles. Community rights or collective rights have also been bestowed in addition to individual rights. Each family became entitled to a maximum of four hectares of forest land. These rights have been conferred without following the process of the FCA. Subsequently rules were framed under the FRA which laid down procedures for implementing the Act. However, the implementation of the Act has been a story of a mixed successes and failures. Till May 2015, the government had received 4.4 million claims, but only 1.7 million titles were approved. The community forest rights are still lagging behind (Oommen 2015). One successful example is that in Gadchiroli district of Maharashtra, about a thousand villages got community forest-rights titles and earned millions from sale of tendu leaves and bamboo. The process under the FRA still remains unfinished. There have been divergent views among stakeholders. A dozen activist NGOs have launched a 'campaign for survival and dignity' in 10 states. They support the Act but criticize its poor implementation.

There are groups of conservationists who oppose the FRA as they argue that its implementation will harm forests and wildlife. A petition filed by some conservationists with the plea that the FRA was a land distribution scheme, and as such not within the jurisdiction of the GOI to make laws on, was dismissed by the Supreme Court of India on 31 July 2007.

Over the last three decades, fragmentation of the habitat has been identified as the single largest threat to biodiversity. The FRA has set the stage for another round of massive fragmentation, which will lead to serious human–wildlife conflict (Bhargava 2011). No other law relating to the forests is as controversial or contentious as the FRA. It has created confusion within the forest departments, revenue department and civil society on what is and what is not allowable under this law. A study concluded that only a small fraction of tribal and other households have been benefitted by the FRA (IGS 2013). The effect of the FRA on deforestation has been negligible. There was some clearing of forest areas too by the claimants of land right to inflate their claim for more area than was in records. This study was, however, confined to a small area in the Udaipur district of Rajasthan state. Whereas individual rights have been conferred in thousands of cases, the progress with community rights over the forest lands has been limited.

Speaking to *Down to Earth*, an environmental magazine, N. C. Saxena, a retired bureaucrat who headed the National Committee on Forest Rights Act, said, 'Transferring the power to communities without their claiming right may not work because forest management depends on how "cohesive and capable" the communities are, and in such situations joint forest management should continue as an interim measure till the forest department withdraws and JFM becomes community forest management' (*Down to Earth* 2017). The issue of FRA implementation is likely to continue to remain volatile and an unfinished agenda in coming years, as there will be nothing like a final settlement. The FRA opened up a floodgate and thousands of fake claims were filed for land grabbing, most of which were rejected. It will be difficult to address all existing and new issues and, like other controversies of a similar nature, it will not see closure. This also reflects contradictions and conflicts across various departments of the government and stakeholder communities. However, in a democratic form of governance, these situations are inevitable and one has to learn from these generation after generation and then seek direction from the judiciary about the real implementation of the law and its interpretations when the executive fails to have an impartial and objective understanding of the vision and intents as well as the desired outcomes.

Historical wrongs are impossible to set right across the globe, and in making such an attempt, disagreements and conflicts are bound to take place. Whatever has been done cannot be undone, and what was the situation before colonial occupation cannot be restored today. Some NGOs have forecast a more radical struggle on forest rights in the future which will challenge the environment–development boundaries.

Judicial Activism

There were no well-planned and concerted policy implementation efforts by the central or the state governments, and while the policy was in place, deforestation, forest depletion and destruction continued growing at an unprecedented rate. This depletion still continues in most parts of the country. There is an illusory perception among forest officials that the forest depletion has slowed down, that the fuelwood collection from forests and grazing has reduced and that timber thefts are within control. But this opinion fails to explain the cumulative existence of vast stretches of degraded forests.

When the central government, state governments and the SFDs were failing at implementing the forest policy, the Supreme Court intervened through a writ petition filed by T. N. Godavarman Thirumulpad in 1995. Another writ petition was filed the same year (1995) by the Centre for Environmental Law, seeking judicial intervention for implementation of the Wildlife (Protection) Act 1972 (WP 337 of 1995) for determination of rights in protected areas. These two writ petitions, being public-interest litigations, were extended in scope and all states were brought into their purview. Since then, the Supreme Court has been issuing far-reaching and significant directions from time to time to both the central as well as the state governments, and thousands of interlocutory applications have been filed and adjudicated.

This judicial activism, as it is generally called, that involves public-interest litigation on matters concerning people at large, is a consequence of the failure of the executive arm of the government to implement its own policies and laws. It is a reflection of the government's failure, despite policies and laws, to prevent forest and biodiversity destruction. A large number of orders relate to the amended FCA of 1988, the FCA of 1980 and the Wildlife (Protection) Act of 1972, and the rules and regulations made thereunder have been made by the Supreme Court. The net impact of this judicial activism has been to put a curb on indiscriminate forest harvesting, which was progressing at a fast-increasing pace. The Supreme Court reminded the government to act in accordance with its own policies and laws. Until then, political expediency and an unscrupulous bureaucracy were indulging in a destructive harvesting of forest resources and the conversion of forests to other land uses.

The Supreme Court order of 12 December 1996, for example, directed the effective, full and genuine implementation of the FCA of 1980, as amended in 1988. The court, through another order passed in response to the Centre for Environment Law petition, restrained de-reservation of reserved forests in PAs—national parks and wildlife sanctuaries. This power has now been taken over by the Supreme Court, after a proposal by the standing committee of the National Board for Wildlife was approved. The Supreme Court's landmark orders included a definition of 'forest', something that had not existed in policies or laws before; stoppage of activities such as the running of saw mills, veneer factories or plywood mills; stoppage of mining of minerals in forests without prior approval of the Supreme Court or the central government; a total ban on cutting of tress in the Tirap and Changlang districts of Arunachal Pradesh; suspension of movement of timber from the seven North-Eastern states; a ban on felling of green trees in natural forests—unless dead, dry, uprooted, damaged or diseased—in the hills of Jammu and Kashmir, Himachal Pradesh and Uttarakhand states; the constitution of a high-powered committee to oversee the implementation of the Supreme Court's orders relating to the North-East region; a ban on cutting of *Acacia catechu* trees in Jammu and Kashmir; some orders related to mining in Uttar Pradesh and Uttarakhand, Karnataka, Goa and Madhya Pradesh; a ban on removal of any dead, dry or fallen trees from a PA; compensatory afforestation for diversion of forest land to non-forest uses; encroachment; and so on.

There has been a far-reaching impact of the Supreme Court directives on India's forest management as well. The MOEFCC became more careful in considering and approving forest-land conversion proposals received from state governments. Despite this, certain proposals—like the one for the Posco plant in Odisha and another for coal mining in Jharkhand—came in for judicial scrutiny. Even the Central Bureau for Investigation (CBI) got involved in a few cases where kickbacks were suspected. The issue of compensation for loss of forests and forest lands also underwent judicial review. Also, the use of compensatory afforestation funds and the net present value (NPV) recovered from user agencies in lieu of forest lands released for development activity have been subjected to judicial pronouncements.

The executive tends to exercise arbitrary and even absolute power and those in the higher echelons of the politico-bureaucratic hierarchy resent the curtailment of their executive or administrative and financial powers. Many conscientious forest officers who resisted pressures from those quarters to allow arbitrary and unjustified forest-land conversion, recruitment of staff, allocation of budgets, promotion and posting without following set rules, procedures and court directives were penalized by being transferred to insignificant positions and locations. Even some heads of SFDs suffered humiliation for saying no to unreasonable diktats. Yet a few bold officers took a stand, even in the face of displeasure of powerful people. However, the majority of the officials conformed and have always been ready to please those who hold the reins. The latter claim to be in the mainstream and enjoy all the perks and privileges of their position, and their sycophancy has placed them on a trajectory to success.

The far-reaching impact of the Supreme Court directives has also been seen in the Himalayan states, where the organized extraction of timber was brought down to 10 per cent from pre-1996 levels. The state forest corporations, which had replaced the big timber concessionaires who got forest coupes on royalty for logging of marked trees, became redundant. They were carrying out salvage and extraction of dead, dry, damaged and diseased trees as well as trees that need to be removed to clear forest lands for non-forest activities. The ban on commercial felling of green trees has, on the other hand, certainly resulted in increased illegal cutting of trees by local people and the timber mafia, with or without the collusion of the forest-department officials at the field level, to meet the rising demand for house construction. Whereas the harvesting in natural forests remains suspended, that in the plantations and agro- and social-forestry spaces is permissible.

Because of the Supreme Court directives restricting commercial timber harvest in forests, what used to be 'business as usual' came to a halt. The SFDs discontinued the classical forest management, a legacy of the British colonial empire. They even felt so frustrated that they abandoned preparation of management plans (working plans) in many states. As the working plans, as the name suggests, had traditionally been instruments for cutting trees for more than a century, the forest officials found no incentive to prepare working plans any more. In many states (for example, Andhra Pradesh), the working plan units were either wound up or became defunct without staff postings. The Supreme Court later issued a directive that no forest would be managed without a valid working plan, duly approved by the central government. In Andhra Pradesh, the WB project in 2002 included as a condition of credit disbursement that all forest divisions should have working plans. Similarly, the Thirteenth Finance Commission that determines the sharing of revenue between the centre and the states puts the preparation of working plans as a condition for a state to receive central grants, referring to the Supreme Court orders and stipulations on the 1988 policy (paragraph 4.3).[1] All this prompted a revival of the working-plan culture once again, albeit with a lower commitment and hesitancy. This also suggests that the forestry establishment on their own do not act and function in a way that would help sectoral development and exhibits a clear picture of weariness and apathy.

In summary, the Supreme Court directives halted the fast-increasing devastation of the forests which has been continuing since Independence, involving a nexus among unscrupulous contractors, politicians and the forest bureaucracy. The destruction in North-East India was brought to a halt, and elsewhere slowed down. Interestingly, the Forest Survey of India showed an increase in the forest cover

---

[1] See https://dea.gov.in/sites/default/files/Guidelines%20for%20Forest.pdf (accessed on 14 November 2018).

between the period of 1998 and 2010, some of which could be safely attributed to judicial activism. However, according to Dutta (2007, p. 590), the judiciary at best can only supplement the efforts of the state and its machinery in enforcing the law and cannot be its replacement. He further goes on to say that in the face of a non-responsive administration and a disinterested political leadership (or one having vested interests), it is very unlikely that the orders of the court will be obeyed.

The judiciary is still active in forest conservation. The National Green Tribunal is also adjudicating forest- and environment-related cases. The Central Empowered Committee (CEC), set up under the orders of the Supreme Court, is assisting the latter in disposing of hundreds of interlocutory applications arising from the T. N. Godavarman Thirumulpad vs the Union of India case. Development projects in the PAs have generally been discouraged and approved only after judicial scrutiny, although recently aberrant trends are setting in.

CHAPTER 10

An Assessment of India's Forest Governance

Governance is not 'government', nor it is synonymous with administration. The United Nations Development Programme (UNDP) has defined governance as 'the exercise of economic, political and administrative authority to manage a country's affairs at all levels. It comprises the mechanisms, processes and institutions through which citizens and groups articulate their interests, exercise their legal rights, meet their obligations and mediate their differences' (UNDP 2006). Various authors and organizations have produced a wide range of definitions. Some are so broad that they cover almost anything, such as the definition of 'rules, enforcement mechanisms, and organizations' offered by the World Bank's (WB's) *World Development Report 2002: Building Institutions for Markets* (WB 2002). Others focus more closely on public sector management issues, including the definition proposed by the WB in 1992—'the manner in which power is exercised in the management of a country's economic and social resources for development'.

According to the WB, 'governance' refers to the manner in which public officials and institutions acquire and exercise the authority to shape public policy and provide public goods and services. (Corruption is one outcome of poor governance, involving the abuse of public office for private gains.) The WB came to realize that most of the crises in developing countries are related to governance. Hence the contemporary adjustment package emphasizes governance issues such as transparency, accountability and judicial reform. In this context, the WB has introduced a new way of looking at governance—'good governance' (World Bank 1993a).

Forest Governance

There is no single definition of 'forest governance' either. Also, it is difficult to define forest governance. When we talk of assessing it, we are faced with a question as to what exactly should be assessed. Generally, it is believed that governance is relegated to the government. However, it includes much more than the government, since a wide range of public and private actors makes decisions about forests. Forest governance relates to the processes that are followed to make decisions about forest management, but does not encompass what those decisions are or their outcomes. It is highly complex, however, as it involves multiple stakeholders, services and products. Often, the complexity of the forest sector results in conflicts and conflict of interests. There is a never-ending conflict for forest land and for forest goods, such as wood and non-wood forest products (NWFP). The forests constitute a natural resource that has traditionally been considered, in developing countries, as a common property subjected to overuse, leading to degradation and deforestation. Reluctance and failure to address the relevant issues and resolve conflicts is a reflection of poor governance. Non-implementation of policies and laws may lead to an anarchic situation that would be a net outcome of poor governance.

Challenges of Forest Governance

The most common conflict among stakeholders is for land and products. This causes overuse of resources, degradation and deforestation. At the same time, worldwide, forestry remains a low-priority sector with little or no political clout. For example, in India, forests have the semblance of *de facto* common property resource when it comes to its use for pasture and fuel wood collection. The forests in tropical countries inevitably have suffered from the tragedy of commons. The institutions involved with forest resource have a very complex relationship, giving rise to a complex governance structure. Various stakeholders or institutions that influence decision making in forest sector are depicted in Figures 10.1 and 10.2.

Forest governance is not independent; it is part of a system. Its quality and effectiveness reflect a country's overall governance system. Achievement of forest policy goals and enforcement of forest laws or sectoral goals are not possible unless these are supported by the system. The 'system' is the country's overall governance web, which is highly intricate. In India, both state and central governments play key roles by promulgating policies and laws, though the states implement policies and enforce most laws. Therefore, an assessment of forest governance will be required to be done at both levels. The quality of governance determines whether forest resources are used efficiently,

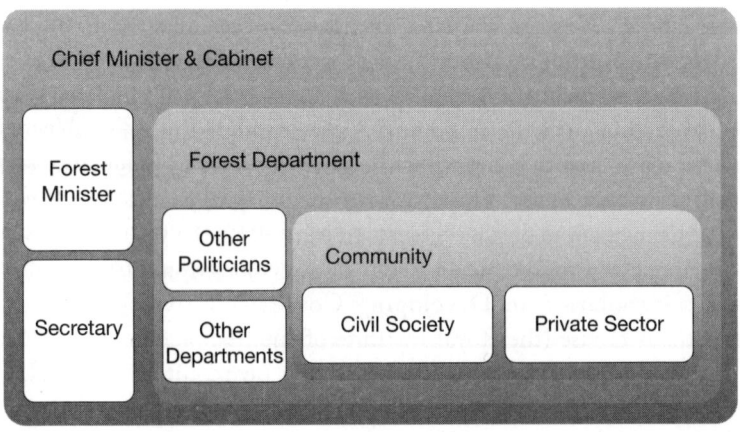

Figure 10.1 *Stakeholders That Influence Decision-Making in the Forest Sector*

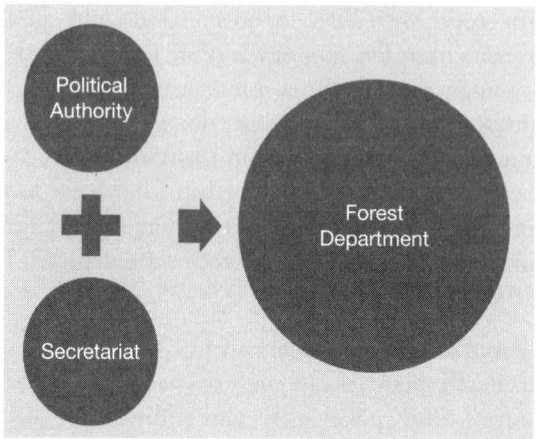

Figure 10.2 *Forest Departments in the State Government Set-up*

sustainably and equitably, and whether a country can achieve forest-related development goals.

Development of an Assessment Framework

In 2009, a number of international organizations initiated meetings and discussions focused on forest governance. In February 2010, the European Union (EU) organized a meeting on Forest Law Enforcement, Governance and Trade (FLEGT) in Rome, in collaboration with the Food and Agriculture Organization (FAO), where participants recognized the need to develop pragmatic and feasible indicators for forest governance and agreed to hold an international workshop on this subject. In May 2010, the United Nations Collaborative Programme on Reducing Emissions from Deforestation and Degradation in Developing Countries (UN-REDD) and Chatham House (the Royal Institute of International Affairs of the United Kingdom) organized a workshop on monitoring REDD+ governance safeguards. In September 2010, the WB, the FAO and the Swedish International Development Agency (SIDA) organized a workshop in Stockholm on development and the application of indicators for assessment of forest governance. It was decided to develop a core set of principles and criteria of a generic nature, which could be flexible enough to apply in different country settings. The WB-led Program on Forests (PROFOR) and the FAO were called upon to undertake this task. The framework was further developed by these agencies in close coordination with UN-REDD and the Chatham House initiative.

The framework so developed has been used for the description, diagnosis, monitoring, assessment and reporting of the state of forest governance in a country. It can also assist in stakeholders' consultations on forest-sector governance as well as improving governance. According to the FAO (2011) the framework draws on several approaches currently in use or under development for the important processes and initiatives relating to forest governance. It includes the WB's forest-governance framework for reforms; the World Resources Institute's (WRI) Governance of Forests Initiative (GFI); the criteria and indicators for sustainable forest management; and the proposed draft of the Monitoring Governance for Implementation of REDD+. The WRI used its framework in Brazil, Indonesia and Cameroon to assess forest governance for REDD+. A requirement of the REDD+ international agreement as part of the Paris climate agreement is that governance issues should be addressed while developing a national REDD+ strategy.

Pillars and Principles of 'Good' Forest Governance

Good governance

A related concept and commonly used phrase is 'good governance'. Again, it would be difficult to define it. There is a general consensus that it is associated with principles such as transparency, participation and accountability, as well as with effectiveness and efficiency of managing natural, human and financial resources with equity (rule of law). Good governance is a critical foundation for achieving positive social, environmental and economic outcomes.

The FAO and PROFOR (the World Bank) forest governance framework is based on three pillars and six principles of 'good' forest governance, and it builds on the principle that governance is both the context and the product of the interaction of a range of actors and stakeholders with diverse interests. The following three pillars of the framework are fundamental to forest governance (FAO 2011):

(a) Policy, legal, institutional and regulatory frameworks
(b) Planning and decision-making processes
(c) Implementation, enforcement and compliance

Pillar 1: Policy, legal, institutional and regulatory frameworks

This pillar includes the policies and legal framework related to the forest sector and how these elements of other sectors affect the management and conservation of forests. This pillar also examines the clarity and comprehensiveness of these frameworks.

Pillar 2: Planning and decision-making processes

This includes accountability, transparency and inclusiveness in key forest-sector institutions and governance. This pillar examines the effectiveness of the processes and institutions and how key organizations function, how they enable stakeholders' participation and how accountability in decision-making by executive authorities is ensured.

Pillar 3: Implementation, enforcement and compliance

This pillar explores how effectively, efficiently and equitably regulatory and institutional frameworks are implemented.

Components

The FAO (2011) gives a comprehensive description of the components and sub-components of the three pillars. There are 13 components and 80 sub-components. The assessment framework also provides guidance on how the indicators can be developed to assess the elements of each of the pillars. The indicators may be of any number, not pre-determined and should be country- and circumstance-specific. The WRI identifies 122 indicators of forest governance.

Who Should Assess Forest Governance

An assessment of forest governance in all the states as well as the central MOEFCC in India is a highly desirable and necessary step, as the documentation can be used for undertaking important reforms in the forestry sector and institutions. Not that everyone working in the forestry sector has to learn skills to evaluate the quality of existing governance. An assessment of forest governance should invariably be done through an independent agency for objectivity and credibility. The agency should facilitate and not judge. However, many bureaucrats believe that they already know the quality of governance in forest sector and need not be told. There is no doubt that those in the public forestry institutions understand the strengths and weaknesses of governance and that the weaknesses point to the lack of good governance, though they have not themselves or through an independent agency got the governance assessed. Their experiences and perceptions give them a fairly good understanding. Therefore, in any governance-assessment exercise, the officials of these institutions would play a key role—without their participation, a reasonably realistic assessment is impossible. Thus, interactive workshops, where indicators are developed in a participatory manner and are used in the evaluation of governance in a participatory way, are the best way forward. What are the procedures and processes which the public servants follow in taking decisions for implementing policies and programmes and for providing common goods and services to people? This is what constitutes governance. Poor governance gives rise to corruption and the denial of people's legitimate dues.

India's Forest Governance Assessment

This framework discussed above can be used for a rapid and general assessment of India's forest governance that will vary from state to state (see Table 10.1).

Table 10.1 Assessment of Forest Governance in India

	Pillars / Principles	Existence of policy, legal and institutional frameworks	Planning and decision-making processes	Implementation, enforcement and compliance	Scoring on a scale of 1-5 (1 = poor; 5 = excellent)
1	Accountability	Policy, legal and institutional frameworks exist but could have greater accountability	Planning and decision-making processes exist, including preparation of management plans	Inadequate	3
2	Effectiveness	Implementation and enforcement are inadequate	Partially effective	Partially effective	2.5
3	Efficiency	Overall efficiency is at desired level	Working-plan preparation generally shows low efficiency	Inadequate	2.5
4	Fairness/equity	Framework is fair and equitable	Partially or marginally effective	Marginally satisfactory	3

5	**Participation**	Participatory approaches are in place but are not sustainable	In village level micro-plans for forest management, there is community participation in planning; however, planning is confined to a small cell, where even internal participation is very little	Occasionally good participation when joint forest management (JFM) is actively implemented, but is often not sustainable	2
6	**Transparency**	Policies and laws are in public domain; however, disclosure of information to the public is minimal	Information is not disclosed to the public; transparency is not observed in the planning and decision-making process	It is an opaque system by design—the public, including researchers and academics, do not have access to information as to how the institutions are complying with their own rules and procedures and how these are being enforced	1
	SCORE on a scale of 1-5 (1 = poor; 5 = excellent)	4	2	2	Average: 2.5

Accountability

The procedure, processes and regulatory framework seek to ensure accountability through audits and a system of vigilance (both internal and external), a system of hierarchy, checks and balances, reviews and staff performance appraisals. However, if illegal tree cutting takes place or forest land is encroached upon due to poor enforcement of law and compliance, generally even if the officials don't take action, they are not held accountable. The action involving enforcement and compliance may not be at all effective—which, for example, can be observed in the case of evictions from encroached land or in the prosecution and conviction of offenders. In case the forest department officials manage nurseries and plant trees, they are generally not held accountable if there are failures in the plantations or they are damaged due to grazing or fire. When officials are transferred, the new incumbents often do not take up the responsibility of the previous officials' work and are not held responsible for lapses on the part of their predecessors.

Poor accountability also results in the use of obsolete and poor technology and technical packages that do not ensure the desired level of success. Deforestation and forest degradation have been going on for more than a hundred years, but have never involved any questioning of the managers whose planning, actions and management have caused such serious forest degradation and deforestation. The managers never find themselves answering the stakeholders' questions about loss of forest lands and biodiversity. They never believe that they are under any obligation to share information with the public. The websites of the forestry institutions have little useful information. There is, on the contrary, a tendency to hide information so that accountabilities are not fixed. These websites are not updated regularly. If information is required, a citizen has to move the courts under the Right to Information Act (RTI). The information sought may sometimes be denied or provided selectively. One reason for the lack of transparency is that society at large is indifferent to what is happening with the forests and does not put this sector under scrutiny to find out how this national asset is being managed. The lack of concern does not encourage the institutions to share information on their successes and failures. Across the world, the bureaucracy generally is secretive and believes that it alone can take care of national interests and that a lay person would not have adequate understanding of the intricacies involved in the governance and management of natural resources. Sometimes the media publishes news about sensational events or the detection of some corruption, but there is no regular media coverage or probity in a systematic manner.

Two consequences of lack of accountability and transparency are malpractices and corruption. Whether action is taken against those who are held accountable remains a question. How accountabilities are fixed, how public servants wriggle out of taking action and how failure and incompetency are concealed are matters that are shown little concern. The main issue is that the public servants are brazenly unaccountable to society and, if any, the accountabilities are internal to satisfy institutional regulations and procedures.

There is no doubt that under government rules and regulations, the accountabilities of public servants are well laid down and any actions and decisions on their part are subject to review and scrutiny by internal as well as external agencies such as the comptroller and auditor general of India. Anyone contravening the rules and regulations and committing violations may invoke action, which may lead to punishment. There are deterrents in place that safeguard against massive scandals. However, it is also a fact that only one out of a hundred cases is reported (and then action taken), with ninety-nine going undetected and with impunity.

When it comes to the effectiveness and success of projects and programmes involving expenditure on field operations, accountability becomes limited to accounting procedures and the maintenance of accounting books and other papers. For example, no one is held accountable for the success or failure of a plantation, damage to a plantation due to grazing or vandalism, huge mortality, or the use of poor-quality nursery seedlings and resultant poor plantations. Historically, the forestry institutions have not been held accountable for the absence of regeneration or reforestation after heavy and excessive harvesting of trees or unauthorized grabbing of forest land. Generally, a system of checks and balances should be in place. Financial accountability is, by and large, observed by way of bookkeeping and financial procedures and codes are followed. These are subjected to audit also, and if any violation is detected, any irregularity causing a loss to the state exchequer seen or corruption suspected, action under the law becomes inevitable. Major financial irregularities may be subjected to investigation by anti-corruption agencies and may lead to prosecution and conviction of the responsible official. This keeps a check on arbitrary decision-making.

Mismanagement of Natural Resources

The SFDs are custodians or trustees of the forests. That means that they are holding in trust the natural resources which belong to the people, for maintenance, management, conservation and for providing goods and services to the people from the forests. However, this dogma does not find any conviction and the trustees act like feudal landlords. Some people sarcastically say that the forest departments are the last and biggest landlords left in the country. Failure to manage the forests efficiently, as evident from the depleted state of forests, is a breach of the public trust and a clear case of historical blunder, whatever the rationale or argument put forward. All kinds of natural resources, such as minerals and land, have been exploited from time to time, in many cases for private gains. Allowing the illegal cutting of trees and illegal clearance of forests for agriculture or other uses are two such activities involving corruption. It is also a fact that a vast majority of the forest officials have fervently remained engaged in protecting the forests, even at times endangering their lives. Some have even lost their lives.

Transparency

Since the intentions of the public institutions in this sector are not to engage the public by voluntarily providing information on activities and programmes or their implementation and outcomes, transparency is not respected or encouraged. This state of affairs is a legacy of the colonial empire and the post-1947 feudal attitude, which encouraged public institutions to make all efforts to ensure that only selected information was shared, those relating to populist programmes that could serve as publicity material for political interests. The little information that is put out as public disclosure is not at all comprehensive enough to be useful to stakeholders. For example, on the websites of the SFDs, there is no financial information on the development schemes and projects, on the success or failure of plantations or on illegal cutting of trees or poaching of wild animals.

Lack of information is a major hindrance for researchers and for civil society, the very people who can provide timely feedback and valuable inputs to the government for improving policies and programmes. The public institutions are markedly reluctant to and inactive in disseminating information to stakeholders. For example, stakeholders would like to know how the existing JFM programmes are doing, the number of village-level institutions still active in participatory forest management (PFM), the number of functional and dysfunctional community-based organizations, investments in JFM/PFM, the sustainability of the programme and also rural institutions (village councils) involved in this programme. Lack of transparency is also partially due to the indifference of the stakeholders to know about* and seek information on sectoral performance.

Lack of effective communication, both internal and external, is a result of the tendency to monopolize information and not share it with others due to a sense of insecurity and the belief that information and knowledge are power. The political system also discourages dissemination of information. It is the RTI Act that has brought the compulsion to divulge information on demand. However, volunteering information is not a part of the established governance framework. This is evident in the fact that, while drafting replies to the parliament or state legislature questions, civil servants tend to provide the minimum information and it is a practice not to volunteer information that may lead to inconvenient supplementary questions. Dissemination is therefore selective, that too for positive publicity, and in some cases, for raising awareness on certain aspects of a programme or scheme.

Transfer and Postings

The frequent transfers and postings of SFD officials at all levels of organization has made them poorly accountable, complacent and frustrated. They lack the desirable level of commitment and enthusiasm, traits that were common before 1970. The Indian Forest Service (IFS) was constituted as an all-India service in 1966 with lot of expectations, but has not been successful in delivering what it was supposed to. It lacks allies and support from bureaucracy, politicians, civil society and the general public, which has made it an inward-looking and demoralized institution, in which each individual member is focused on his or her personal posting, promotion and career advancement.

There is opacity in cases where arbitrary powers are used at different levels of authority. One of the most misused arbitrary powers is the transfer and posting of government employees. The transfers are done in a large number of cases not on the basis of merit or in the public interest, to improve the administration and management, but to favour some at the cost of others who are unable to mobilize support or recommendations from powerful personalities. These actions of the authorities cannot be challenged in a court of law because, first, the courts generally take the view that they should not interfere in administrative matters and, second, a victim may be further victimized if he or she challenges a wrongful transfer. Instead, those victims who are shunted to insignificant positions and inconvenient locations have to tolerate and suffer indignity, trauma and demoralization. It is a well-known fact that transfers are used as an instrument for penalizing some employees if they earn the displeasure of the superior authority or do not oblige the latter by refusing to perform an unreasonable task; then the latter seeks vengeance. Transfers and punishment postings send a message across that others may face similar situations if they act in sheer disregard of the wishes of those in power. Besides, there are allegations from time to time that kickbacks, bribes and even bidding play a role in transfers, postings and even promotions. In the states, in particular, the politicians have an obsessive addiction to ordering frequent transfers that keep the officials on tenterhooks, compelling them to spend a significant part of their time and energy in maintaining their positions as long as they can and yet continuing to suffer from a depressive sense of insecurity. The short tenures of officers and staff in responsible positions make accountability scarce. All this makes governance poor.

Corruption

In the forest sector, corruption is found in the procurement of goods and works. In procurement, though due diligence is observed by following a rigorous bidding process and procurement decisions by committees, yet some misprocurements do take place. If a certain bidder is favoured or a bidder manipulates the process with or without official connivance, a doubt could be raised about the integrity of the process. A large number of suppliers tend to influence the procurement process by kickbacks to those who matter at the decision-making and approval levels. And, as greed is a human tendency, there is no dearth of public servants at different levels who fall prey to it. Decentralized petty procurement may also involve corruption.

In departmental operations, when the field staff hire labour, skilled or unskilled, for harvesting and planting, the wage disbursement is generally inflated by adding a fake number of labour days when there was no actual labour engagement. The difference between wage disbursement and the expenditure actually booked goes into the pockets of field supervisors and part of it may even travel upward (also see Box 10.1).

In all government departments, when senior officials tour, the local staff make arrangements for hospitality—food, drinks and other comforts—and the expenditure incurred is rarely contributed by staff out of their salaries or from a government-approved budget. It comes from illegal earnings, contractors or fudging accounts. The public works department, power department, revenue department and indeed all other departments have this 'tradition' that must be respected, as the unwritten rule goes. In some cases, allegations have been made regarding the handling of public auctions of forest products. The state forest corporations (public-owned companies) that harvest wood from the forests are generally alleged to adopt corrupt practices in the sales of wood directly or through public auctions. The contractors awarded work are also expected to please the officials, who inspect their work and approve payments. In some departments, the kickback amounts or percentages are fixed and are built into quotations given in the tender or bid documents. Ultimately the payoffs come from public money or taxpayers' money.

Box 10.1 Integrity

There is a serious issue of integrity, not only financial but also administrative. Bureaucrats have an irresistible tendency to manipulate, distribute and even fudge statistical data to prove a point or support a favourable piece of information. For example, the data on afforestation and tree planting during 1980–1990 lacked integrity, as it did not reflect the ground realities and supported inflated numbers to show that the set targets were nearly achieved. Another issue is that in many cases, an unrealistic target is set and then efforts are made to show that it has been achieved by relying on anecdotes and conjecture. The government tends to give a rosy picture of every programme and project. However, when implemented on the ground, the outcomes do not always match the intentions. Moreover, all monitoring systems measure and report inputs and outputs, but hardly any genuine efforts are made to evaluate outcomes and impacts.

The most gruesome example is that on 1 January 1985, the Prime Minister of India announced—as advised by bureaucrats and some NGO activists—that 5 million hectares of wasteland would be afforested every year. To undertake this ambitious programme, the National Wasteland Development Board (NWDB) was set up under the chairmanship of an NGO activist, with a secretariat and a number of civil servants. The NWDB worked to achieve that target but failed. It made an all-round effort to force state governments and their forest departments to accept an enhanced annual tree-planting target. The problem that the states faced was that of finance and capacity. Despite allocation of funds from rural development schemes and huge external assistance, this target could never be achieved. During the five years from 1985 to 1990, an area of 8.86 million hectares was planted, against the target of 25 million hectares. It is another matter of speculation how much of the coverage was actually achieved and how much survived after a few years on the ground.

Target-driven programmes affect both quantity and quality. JFM is another example, where field staff were forced to organize village-level forest committees to enable rural communities to participate in this state-administered programme. Coming to yet another national programme, the Green India Mission (GIM) set an ambitious target of 10 million hectares to be covered with plantations, both in and outside state-owned forests, over a period of 10 years. It is going to meet the same fate as wasteland development. The bureaucrats set a target to make good grounds for publicity and propaganda and to please their political masters, without considering the constraints. When it comes to mobilizing resources, they fail miserably. Unrealistic assumptions lead to unrealistic expectations.

India's nationally determined contribution (INDC) submitted in 2015 to the UN as a pledge to implement the Paris Agreement on climate change is another example, as it relates to the forest sector's contribution to carbon sequestration. India has made a commitment at the international level that additional forests will be created to develop sinks for 2 billion mtCO2e (metric tonnes of carbon dioxide equivalent) annually by 2030. An estimate of the financial resources needed for this is ₹10,000 billion (equivalent to US$1600 billion). The question is why such an unrealistic target was set and why credibility is being compromised nationally and internationally. This is also an indication of bureaucratic illusions, reflecting on professional integrity and ethics.

The taxpayers' money is usurped by corrupt employees holding public offices at various levels in the government hierarchy. Despite all this, there is a large number of government officials who maintain their integrity, albeit they constitute a minority in the whole complex system. Corruption is termed by corrupt officials as an 'incentive' that comes with a position; they would not work without such incentives. That is why some posts in the government are termed plum or lucrative posts and those providing no opportunity of 'incentives' are dry posts. That is one reason that there is always fierce competition for postings among government employees, further exacerbating corruption. Corruptions has been so deeply entrenched in Indian society that it is not associated with any shame or guilt and its eradication in the foreseeable future is impossible. At the most, it could be reduced by discouraging corruption at the political level.

A strange phenomenon that is encountered is that corruption is carried out so shrewdly, surreptitiously and with mutual collaboration and unity among all parties that very few cases are reported or detected, investigated or prosecuted. It is just like a food chain in which everyone gets his or her share (or cut) and all are happy in the system.

Public dealings too provide an opportunity for unlawful money-making, which involves government public utility and municipality employees receiving and processing applications for services and grievances to ensure redressals for citizens. By delaying or deliberately harassing service-seeking citizens, those interfacing with them compel people to pay bribes. Many public-utility departments and regulatory agencies such as municipalities or tax authorities are alleged to have indulged in harassing common citizens and corrupt practices. Here the officials demonstrate and exercise their powers as enshrined in the relevant rules. In the forestry sector, applicants may be beneficiaries for getting fuelwood or timber on concessional prices for their bona fide use, it being a traditional right of the concession holder, or may be seeking permission to cut trees on their private land and transport the wood. Small contractors bidding for petty contracts for logging or construction may have to pay bribes. In these and other similar cases, power is demonstrated by denying access, raising objections on non-submission of complete documents, delaying processing of legitimate cases or simply arbitrarily refusing to accept and approve requests. Often these public interactions cause harassment. Ultimately, the officials are hinting for kickbacks. For those who give bribes (often called 'greasing of palms' or 'putting weight into paper'), their matters move fast, and sometimes at lightning speed, depending on the amount of the kickback paid in cash. Generally, the motivation of the public-dealing officials is how to make private gains under the regulatory framework and how to enforce laws that can yield private benefits. They do not believe in public service because even they are public servants, but in practice they are masters. From petty, low-level officials to high-level officials, there is a conviction that they are there to rule and not to serve and that they strongly strive to preserve their feudal tendencies.

No one complains, as the senior persons do not try to redress grievances—and the matter is ultimately passed on to the responsible lower-level functionary who gets angry with the complainant and then makes all efforts to completely destroy the case. It is not uncommon that papers and files are misplaced in government offices (file not traceable) and no one makes a serious effort to recover the lost papers. In many cases, such misplacements are intentional, to harass people seeking redressal of their grievances or legitimate permissions, permits, license, authorization or orders. The system has become so unethical, arbitrary and corrupt and so deeply entrenched in culture that people have forgotten to distinguish between right and wrong. Before Independence and in the 1950s and 1960s, giving a gratuity to lower-level employees in government offices was small cash for tea; it later became a 'bribe' and now it is extortion. Thus no one seems to be accountable to the people and no one cares, because wrongdoings are perpetrated without consequences and the strange reality is that society has accepted this norm.

Many people who are afraid that the government officials may not listen to them. They approach politicians such as ministers or legislators, and bring recommendations from them to officials. One can easily see crowds of people at the bungalows of politicians in the mornings. The politicians sometimes recommend actions that may be contrary to established rules and it is up to the officials to exercise their judgement in such cases. When a person appears before a government official with a recommendation, presuming that he or she has orders from superior authorities, the expectations become so powerful that any refusal is not taken in good spirit. The feedback given to the politician by the individual is generally distorted and generally provokes the politicians.

Scruples have poor existence at various levels among authorities. The reality is that those who are honest and maintain their professional and personal integrity are often victims of whims and fancies and the arbitrariness of powerful superior authorities. They are marginalized and sometimes ostracized by the system and are assigned insignificant positions with little or no work, poor infrastructure and few facilities to work and live with. Integrity becomes for them a disincentive for such people. On the other hand, unscrupulous government officials with their sycophantic behaviour, keep their senior and superior authorities happy and in good humour and everyone in the hierarchy obtains personal gains. There are many exceptions to this generalization but that happens when situations are conducive and convenient to officials with integrity and scruples. But there are few exceptions and business goes as usual.

Part V

Emerging Global Issues

Commitments and Challenges

CHAPTER 11

Climate Change and Forests

India signed the United Nations Framework Convention on Climate Change (UNFCCC) in 1992, along with 192 other countries, as a response to the urgent call to work together to save our planet from the disastrous consequences of climate change being exacerbating due to the increased concentration of greenhouse gases (GHGs) in the atmosphere. Since then, India has been very proactive and vociferous in negotiations on international climate change on such platforms as the Conference of the Parties (COP) to the UNFCCC and its subsidiary bodies and groups. India is also a signatory to the Kyoto Protocol, seeking benefits out of it by offsetting the emissions of the developed countries who are responsible for high levels of emissions since the Industrial Revolution. The Kyoto Protocol failed to be effective in practice due to its non-ratification by the United States of America (USA) and the withdrawal of a few other countries such as Canada and Australia.

The Kyoto Protocol provided financial incentives for offsetting carbon emissions from developed countries under the clean development mechanism (CDM). In the forestry sector, this was to be done through afforestation and/or reforestation (AR). Undoubtedly a small number of AR projects were registered for seeking financial incentives in the voluntary carbon market. The process and procedures, including verification, registration and monitoring, were so complex and cumbersome that most developing countries could not take advantage of AR. The cost of transactions have been so high that the returns were often estimated to be negative.

Climate Change

It is evident from scientific work that climate change is happening. Its impacts are already being experienced and any scepticism about it is no longer tenable. Emissions of GHGs are still increasing, despite the intentions of the international community to the contrary, as enshrined in the UNFCCC. The Fifth Assessment Report (2014) the Intergovernmental Panel on Climate Change (IPCC) provided more evidence that climate change is real and recommended that the international community should take action to avoid disasters resulting from a global rise of temperature. There is a general consensus among the nations that an increase in the global average temperature should not be allowed to go beyond 2 degrees Celsius by 2050. This goal is now enshrined in the Paris Agreement on Climate Change, 2015.

Impact of Climate Change

The impacts of climate change are being experienced more and more across the world in the form of increased temperatures, erratic rainfall, melting of glaciers and polar ice, rising sea levels, and so on. The concentrations of GHGs in the atmosphere have increased and there is warming of the atmosphere and the oceans. The earth's atmospheric temperature has already risen by 0.85 degree Celsius since the Industrial Revolution (IPCC 2013). It is feared that these impacts may assume a catastrophic form in the coming years. At the earth's surface, each decade during the last 30 years has been successively warmer than any preceding decade since 1850.

According to the IPCC, among other natural systems, forest ecosystems are also being affected by regional climate changes, particularly temperature increases. Examples include:

(a) earlier timing of spring events, such as leaf unfolding, bird migration and egg laying
(b) poleward and upward shifts in ranges of plant and animal species
(c) earlier 'greening' of vegetation in the spring, linked to longer thermal growing seasons due to recent warming

Conserving forests and biodiversity will therefore require a two-pronged approach. First, GHG emissions must be radically reduced in order to slow the rate and extent of global climate change. Second, assuming that we can limit the rate and extent of change, we will still need to respond to the change that is already inherent in the system and buy some time for ecosystems while emissions are reduced. We must reduce emissions quickly and deeply and take local action to protect biodiversity by increasing the resistance and resilience of natural systems so that they can better survive the changes to come. If CO_2 emissions are not reduced quickly and deeply, some of the valuable ecosystems will be lost forever.

Emissions from Forests and Other Sectors

Climate change is being caused due to increased concentrations of GHGs in the atmosphere as a result of burning of fossil fuels and thus it has anthropogenic causes. Deforestation emits 17 per cent of the total emissions of carbon dioxide (IPCC 2007, 104), which is more than the transport sector (13 per cent). Forests are cleared across the globe in tropical countries for growing agricultural crops, such as oil palms, rubber, coffee, soy, and so on; for cattle ranching; urbanization; mining; and infrastructure development. According to the Food and Agriculture Organization (FAO) of the United Nations (UN), about 13 million hectares of forests are cleared annually. This destroys the carbon sink which the forests constitute, thereby reducing carbon sequestration by the forests.

Deforestation continues at an alarming rate even today. This is especially happening in tropical countries, with the highest rates of deforestation being in the Amazon basin, Indonesia and the Congo basin. The IPCC, in the report of its Working Group III released in

2014, estimated 24 per cent (12 GtCO2eq) of net emissions are from the agriculture, forestry and other land use (AFOLU) sector.

Climate-Change Mitigation and Adaptation

If deforestation is avoided worldwide, the emission of CO_2 will be reduced and forest conservation and sustainable use will help in sequestering carbon from the atmosphere, thereby helping in reducing GHG concentration in the atmosphere. Reduction and/or prevention of deforestation is an important option for mitigating the impact of climate change by preventing carbon emissions. The mitigation of climate change by protecting forests is more cost-effective than engineering methods that involve carbon capture, carriage and sequestration underground. In the Kyoto Protocol, 1997, afforestation and reforestation were included in the CDM and qualified for issuing carbon emission reductions (CERs) that could be traded in international carbon market. However, with high transaction costs and low carbon price, this almost failed and brought disappointment to a large number of stakeholders.

Vulnerability

India's forests are highly vulnerable to the impacts of climate change. Increased temperature with prolonged periods of drought will hit not only forest growth but also reforestation and afforestation efforts. Recruitment stage is when seedlings germinate naturally in forests. Climate change is likely to result in increased mortality of young recruits. There is likelihood of changes in the species composition as well as distribution. Xerophytic plant species are likely to persist. At the same time, erratic monsoons may bring heavy rainfall, though it is difficult to see if it will offset the massive casualties due to prolonged droughts and high temperatures causing desiccation of soil and vegetation.

Future scenario 2050

At Copenhagen, an accord was signed by more than 50 countries, expressing their resolve to limit temperature increase to 2 degrees

Celsius beyond the pre-industrial period. However, there are opinions that if the current trend of carbon dioxide and GHG emissions continues, the rise of 2 degrees Celsius will happen long before 2050. In fact, the impacts of climate change may assume catastrophic forms. Extreme events will have greater impacts on sectors with closer links to climate, such as water, agriculture and food security, forestry, health and tourism.

International Agreements on Climate Change

United Nations Framework Convention on Climate Change

The UNFCCC was signed in 1992 at Rio de Janeiro, where the Earth Summit took place. It was signed by 148 countries, which ratified it subsequently. It was signed against the backdrop of the realization that climate change had become a reality and that it would result in serious impacts on the human race. More countries joined later. In all, 197 countries have ratified UNFCCC.

Article 2 of the UNFCCC states its main objective, which is 'to stabilize greenhouse gas concentrations in the atmosphere at a level that would prevent dangerous anthropogenic interference with the climate system'.

Kyoto Protocol

The Kyoto Protocol emerged as a major international agreement and an advance in the history of climate-change negotiations. It included targets for 37 industrialized countries and the European Union (EU) to reduce emissions of GHGs by 5 per cent compared to 1990 levels, between 2008 and 2012. The Protocol recognized that developed countries have been principally responsible for the high levels of GHG concentrations in the atmosphere and grouped countries into two categories:

(a) Annex I: Industrialized countries
(b) Non-Annex I: Developing countries

Under the Protocol, mandatory limits on GHG emissions were placed on Annex 1 countries, under the principle of 'common but differentiated responsibilities'. No binding limits were placed on non-Annex I (developing) nations. The Kyoto Protocol also created an international emissions trading (IET) market based on a cap-and-trade system that enabled Annex I countries to trade allowances with other Annex I countries.

Two mechanisms were provided under the Kyoto Protocol:

Joint implementation (JI): emission reduction projects in Annex I countries could generate credits which could then be bought by other Annex I countries and used for compliance under a regulatory cap-and-trade system.

Clean development mechanism (CDM): Annex I countries pay for credits that occur in developing nations by emission reduction. The buyer Annex I country may then use those credits for compliance under the cap-and-trade system.

CDM was the only means by which developing countries participated in the Kyoto markets. This was because the Parties to the Protocol recognized that the cost of GHG mitigation varied significantly between countries and therefore it would be more cost-effective to implement emission reduction projects in countries where the costs were lowest.

REDD+

When the AR projects and the expected financial incentives were found very inconvenient, cumbersome and unattractive under the Kyoto Protocol, the international community agreed to develop a mechanism of financial incentives for reduction of emissions from deforestation and forest degradation—United Nations Collaborative Programme on Reducing Emissions from Deforestation and Degradation in Developing Countries (UN-REDD). This idea was agreed upon in Montreal at the Fifth Session of the Conference of the Parties to the UNFCCC (COP 11) in 2005. In 2007, in Bali, COP 13 added the 'plus' symbol (+) to REDD. The '+' included 'sustainable management of forests, conservation of carbon stocks and enhancement of carbon stocks'. The evolution of REDD+ continued and its basic principles, which were agreed upon at Cancun (COP 16) in 2010, were finally adopted at Warsaw (COP 19) in 2013 as part of the Warsaw Framework for REDD+. The REDD+ approach was also included in the Paris Agreement in 2015.

COP 19: Warsaw 2013

Listed below are some of the highlights of the Warsaw Framework for REDD plus:[1]

(a) The governments agreed in principle on ways to reduce emissions from deforestation and forest degradation—the decisions were the culmination of seven years of work, and their agreement came as a clear breakthrough for action on climate change.
(b) The set of decisions aimed to encourage forest conservation and sustainability, with direct benefits for people who live in and around forests.
(c) The package provided a foundation for the transparency and integrity of actions and clarified the coordination of support.
(d) It established the means for results-based payments if developing countries could demonstrate the protection of forests.

Decisions were taken by the COP in Warsaw on the following seven subjects:

(a) Work programme on results-based finance to progress the full implementation of the REDD+ activities
(b) Coordination of support for the implementation of activities in relation to mitigation actions in the forest sector by developing countries
(c) Modalities for national forest monitoring systems
(d) Reporting on compliance with the safeguards agreed earlier at Cancun (in 2010)

---

[1] http://unfccc.int/key_steps/warsaw_ outcomes/items/8006.php (accessed on 8 October 2018).

(e) Guidelines and procedures for the technical assessment of proposed forest reference emission levels (FERL) and/or forest reference levels (FRL)
(f) Modalities for measurement, reporting and verification
(g) Addressing the drivers of deforestation and forest degradation

It is the 'plus' in REDD+ that should help in reducing emissions from forest degradation directly, by prevention of further degradation and also by reforestation or restoration of forest productivity of these lands, thereby enhancing sequestration and conserving forest carbon stocks. It will also contribute to sustainable management of forests.

India's Policies to Contribute to International Efforts

India's Forest Policy of 1988 is a broad policy statement, which reflects sectoral aspirations as well as overarching goals, which see the forests to be serving the environmental needs of society along with providing benefits to forest-dependent people. According to the policy statement, the principal aim of the Forest Policy is to ensure environmental stability and maintenance of ecological balance, including atmospheric equilibrium; the derivation of direct economic benefits is subordinate to this principal aim. The policy concludes that the 'forests should not be looked upon as a source of revenue'. As the issue of climate change was not discussed at the time when the policy was adopted, there is no direct reference to this aspect. However, the policy objective to maintain environmental stability through forest conservation can be deemed to have included climate-change concerns as well.

National Action Plan on Climate Change

India prepared its National Action Plan on Climate Change (NAPCC) with eight missions, on solar energy, water, agriculture, energy, habitat, the Himalayas, strategic knowledge and forestry. It was adopted and released in June 2008. The action plan emphasized the overriding priority of maintaining high economic growth rates to raise living standards. The plan 'identifies measures that promote development objectives while also yielding co-benefits for addressing climate change effectively'. It says these national measures would be more successful with assistance from developed countries, and pledges that India's per capita GHG emissions 'will at no point exceed that of developed countries even as we pursue our development objectives'. The GIM is one of the missions, aiming at promoting climate-change mitigation by enhancing forest cover in India over a period of 10 years.

Green India Mission

Mission aim

The mission aims to respond to climate change by a combination of adaptation and mitigation measures:

(a) Enhancing carbon sinks in sustainably managed forests
(b) Adaptation of vulnerable species and ecosystems to the changing climate
(c) Adaptation of forest-dependent communities

Mission objectives (MOEF 2011; MOEF n.d.)

(a) Increased forest or tree cover on 5 million hectares of forest or non-forest lands and improved quality of forest cover on another 5 million hectares (a total of 10 million hectares)
(b) Improved ecosystem services, including biodiversity, hydrological services and carbon sequestration as a result of the treatment of 10 million hectares
(c) Increased forest-based livelihood incomes for 3 million forest-dependent households
(d) Enhanced annual CO_2 sequestration of 50–60 million tonnes by the year 2020

For the target of 10 million hectares, a budget outlay of ₹460 billion (equivalent to US$7.2 billion) was projected for 10 years and it was to come from various sources, including a compensatory afforestation fund, centrally sponsored schemes and rural employment programme.

The preparation of a framework for implementation of this ambitious programme took very long and it was approved only in 2014, with a much truncated budget of ₹2,000.

The central government finally approved the National Mission for Green India (GIM) as a centrally sponsored scheme. Of the total allocation of ₹130,000 million envisaged in the 12th Five-Year Plan (2013–2018), the Plan outlay is ₹2000 crores. The source of funding for the scheme would be from the Plan outlay, and convergence with National Rural Employment Guarantee Act (NREGA) activities, Compensatory Afforestation Fund Management and Planning Authority (CAMPA) and the National Afforestation Programme. The sharing pattern for the plan outlay would be 90 per cent from the centre and 10 per cent from the state for the North-Eastern states and 75 per cent from the centre and 25 per cent from the rest of the states. The 13th Finance Commission grant funds were allowed to be counted towards the states' share, to the extent that this is in conformity with the Commission's award. Thus the MOEFCC got only ₹2,000 crore approved for five years. Further, for the year—for example, 2016–2017—there was a meagre budget of ₹450 million only. The MOEFCC acknowledged the following challenges:[2]

Inadequate budget allocation

(a) Additional allocation of ₹628.51 crore is required for FY 2016–2017
(b) Average requirement of the fund is around ₹1,800 crore per annum
(c) Budget estimate of ₹45.01 crore for FY 2016–2017 was grossly insufficient
(d) Committed liability for FY 2016–2017 is ₹97.87 crore

Other challenges

(a) Staff strength
(b) Space for mission staff in one common area

---

[2] http://www.moef.gov.in/sites/default/files/Green%20India%20Mission.pdf (accessed on 8 October 2018).

(c) Multitude of schemes—NAP, CAMPA

(d) Coordination and convergence in decision-making structures

Mission implementation has a decentralized participatory approach, with involvement of grassroots-level organizations in planning, decision-making, implementation and monitoring. The *gram sabha* (village assembly) and the committees mandated by the *gram sabha*, including revamped joint forest management (JFM) committees, will oversee implementation at the village level, a revamped Forest Development Agency (FDA) chaired by an elected representative at the district and division levels, a revamped state forest development agency (SFDA) with a steering committee chaired by a chief secretary and an executive committee chaired by the Principal Chief Conservator of Forests at the state or union territory level. A governing council, chaired by the minister of environment and forests, and a national executive council, chaired by the secretary (environment and forests) and co-chaired by the director general of forests and special secretary, with the mission director as the member secretary, facilitate GIM implementation at the national level. A multidisciplinary team, representing both the government and NGOs, is mandated to facilitate planning and implementation at the level of clusters or landscape units.

GIM has not made any significant progress so far. There is a very serious mismatch between the ambitions and the financial support. The assumption was that financing would come from CAMPA, NREGA and other external sources through convergence. This did not happen as expected and, though the states developed some projects, implementation has been slow and ridden with financial constraints. It is unlikely that the set targets and goals will ever be fully achieved. The unrealistic planning through the bureaucratic approach may give a sense of illusory achievement but becomes a preposterous action. The reality is that while everyone was doubting the ambitious projections without any financial assurance and convergence, the officials went ahead with planning and documentation over more than five years to deliver results that were lower than expected.

India's Readiness to Implement REDD+

India's submissions to UNFCCC and its bodies

The Government of India (GOI) made a number of submissions to the UNFCCC secretariat, relating to various aspects of REDD+. These documents are available on the UNFCCC website. India's national strategy, as communicated in one of the submissions in 2011, aims at enhancing and improving the forest and tree cover to enhance ecosystem services to local communities (fuel, fodder, timber, non-timber forest products, or NTFP, and carbon sequestration). Carbon services from forests and plantations are one of the co-benefits and not the main or sole benefit. India will implement only the '+' part, implying that deforestation and degradation are not an issue.

However, in another submission, it was accepted that both deforestation and degradation are prevalent. The submission further mentioned that GIM, NAP and some other schemes from the agriculture and rural development sector are financing afforestation programmes with an annual target of 2 million hectares per year, which would sequester 2 million tonnes of carbon per year (and after 2020, 20 million tonnes per year). For this effort, India requires US$2 billion (₹12,000 crore) per year and a substantial part of that should come from UNFCCC. The submission goes on to state that India intends to work on the definitions of REDD+ activities, for example, sustainable management of forests. It is committed to ensuring that full and adequate incentives from REDD+ go to the local communities. However, the forests will not be managed for carbon services alone, but for all the ecosystem services. Incentives for carbon services will be an add-on to the benefits that the local communities are already receiving from the forest ecosystems. The capacity of state forest departments (SFDs) will be developed by the Forest Survey of India (FSI). This implies only the capacity to monitor carbon emissions from and sequestration by forests. However, there was no mention of the need for overall capacity building to implement REDD+ activities.

Choice of starting phase

The submission further claimed that India was well prepared to move into the final phase of results-based actions, which will be fully measured, reported and verified. However, initiation of results-based actions would be subject to agreement on the fixing of national forest reference levels as well as procedures and processes for measuring reporting and verification. This conflicts with the UNFCCC decisions that a country seeking results-based finance has to implement all three phases of REDD+. Secondly, without completing phases I and II, moving to the third phase of results-based actions and monitoring, reporting and verification is not possible.

Safeguards

India's submission mentions the following instruments as safeguards:

(a) India intends to ensure that all REDD+ incentives available from international sources will flow fully and adequately to the local communities that participate in forest management, manage the forest resources or are dependent on the forest resources for the sustenance of their livelihood.

(b) Part of the incentives are expected to be invested in conservation and improvement of ecosystem services, such as biodiversity and NTFP.

(c) Local communities would be encouraged to develop micro-plans to incorporate such priorities.

(d) JFM approach would be employed for project implementation.

(e) The Forest Rights Act (FRA), 2006.

Regarding financing for REDD and the enhancement of carbon sequestration/services, India's viewpoint was that it should be market-based. It believes services can be non-market based for conservation of carbon stocks and sustainable management of forests (SMF). This submission reflects a lack of understanding. The idea of the safeguards is that activities that are implemented to further the REDD+ agenda should not have any adverse impact on environment or people, particularly those depending on forests and especially indigenous people and other local communities. It has to be ensured that they are consulted and that there is informed participation in the consultation process. Their rights have to be respected. The guidance on safeguards was agreed at COP 16 (2010).

Submission to UNFCCC on 3 December 2011

The main essence of the submission was that two types of drivers of deforestation and degradation had been identified: (a) planned drivers and (b) unplanned drivers. The planned drivers are those where deforestation is permitted officially under the Forest (Conservation) Act, 1980 (FCA). The unplanned drivers are those on which the government has no control and include encroachment on forest land, illegal tree felling, livestock grazing, fuelwood harvesting, forest fires, disease, insect pests and illegal mining. It is understood that Indian delegates were proactive and even vociferous on REDD during the climate-change negotiations at Bali in 2007 and, indeed, it was India that, along with other developing countries, pushed the '+' part to REDD. It continued to remain proactive in the two successive sessions of the COP from Copenhagen to Cancun (2010). However, the proactive stance did not translate into domestic action and there remained a conspicuous disconnect between the negotiations and adequate follow-up action at home to work on the plans for REDD plus implementation. This is evident from the fact that during the last eight years, no perceptible action has been taken on initiating REDD+ action.

For REDD+ activities, the government is required to take action to identify the drivers of deforestation and forest degradation, identify activities to be undertaken to address the drivers and develop national- and state-level strategies consistent with it. While developing national or state strategies, issues relating to forest governance, land tenure, gender and safeguards would be taken into account.

Some documents circulated by the Indian government during UNFCCC conferences claimed that India's forests are net sinks and remove 11.25 per cent of India's GHGs from the atmosphere (ICFRE

2009, p. 16; MOEF 2009). However, there is no consensus on this claim, as many believe that India's forests are net emitters of GHGs, as they suffer from serious progressive degradation.

Deforestation and Forest Degradation in India

Deforestation

Deforestation is driven by both official approval under the law for non-forest uses of forests, mainly for infrastructure projects, and also illegally by people who convert forest land for agricultural use or house construction (the latter would be termed encroachment). Official conversion of forests to non-forest uses took place at an average rate of 150,000 hectares per year from 1950 to 1980, and from 1980, when it was regulated under FCA, the rate reduced considerably in the 1980s and 1990s. From 2004 to 2013, an area of 243,000 hectares was cleared under government orders, and in 2013, proposals for the clearance of 330,000 hectares were pending with the central government. According to a news item in the *Times of India* dated 11 June 2013, per

> recent data acquired through RTI from the ministry of environment and forests by a group of environmentalists, the extent of forest land being diverted across the country on an average stands at 135 hectares (around 333 acres) per day. Such diversions are done on various pretexts, say for coal mines, thermal power plants, industrial or river valley projects.

This means that from 1980 to 2013, 1.62 million hectares of forests have been legally cleared (an average annual rate of 49,275 hectares).

India has lost 897,053 hectares of forests between 2000 and 2013 and has gained 253,928.3 hectares, and thus has suffered a net loss of 642,125 hectares. That is, the official loss is an average of 49,394 hectares per year. India's submission to the UNFCCC that implied that deforestation is not a problem was factually incorrect and misleading. More than 2 million hectares are under illegal occupation or have been cleared for agriculture and housing. This shows that deforestation is a serious issue in India, contrary to the general utterances. The only solace is that with the promulgation of the FCA in 1980, the officially authorized clearance of forests has been reduced to one third of the previous decades' annual rate. The main drivers of deforestation are as follows:

(a) Conversion of forest land for agricultural use (legal and illegal)
(b) Clearance for urbanization
(c) Clearance for infrastructure development—roads, railway lines, dams, transmission lines, irrigation canals, water reservoirs, and so on
(d) Mining of ores and minerals (coal, lignite, copper, bauxite, iron, zinc, limestone)
(e) Quarrying of stones
(f) Settlement of displaced persons
(g) Government buildings
(h) Industries

Forest Degradation

Forest degradation is a result of numerous and often complex causes. The consequences are reflected in serious damage to forest ecosystems (which in many locations is beyond recovery), depleted productive capability, scarcity of forest products and further advancement of the degradation process. Degradation ultimately leads to deforestation. Restoration of degraded forest ecosystems is not only difficult but also costly and is a long-duration effort, with many uncertainties and risks associated with the anticipated success. The attempt can restore productivity of the land but the integrity of an ecosystem, once lost, cannot be restored. The current state of forest degradation is a cumulative result of past management of the forests by state agencies, continuously increasing serious human and livestock interference with forest ecosystems, and current threats to conservation. The main causes responsible for forest depletion are as follows:

(a) Heavy exploitation of commercially important species
(b) Commercially oriented forest management, involving clear-felling of mixed forests and their replacement by planting single species
(c) Management with heavy reliance on natural regeneration

(d) Low investment in regeneration or restoration after tree harvesting
(e) Population growth resulting in increased demand for grazing land and forest products
(f) The tragedy of the commons, as use of forests for pasture, fuel-wood and NTFP collection has a semblance of using a common property resource
(g) Poverty forcing people to heavy dependence on:
   i. Wood as a source of domestic energy
   ii. Wood extraction for fuelwood, small cottage-based industries, tea-leaf curing, tobacco curing, restaurants, and so on
   iii. Free and unregulated livestock grazing
   iv. Unsustainable and destructive selective NTFP harvesting
   v. Shifting cultivation
(h) Forest fires, mostly manmade
(i) Low priority given to the forestry sector when making budgetary allocations and investments.

India's forest cover, including trees outside the forests (TOF), store 6,663 million tonnes of carbon (FSI 2013). Reforestation of degraded forests can sequester millions of tonnes of carbon from the atmosphere and store it for decades. The global carbon stored in terrestrial ecosystems (mainly forests, covering 15 billion hectares) is estimated at 2,477 gigatonnes (IPCC 2000). Additional sequestration by degraded forest was about 0.3 tonnes of carbon per hectare per year. It is estimated that land converted to agroforestry in tropical regions could sequester 3.1 tonnes of carbon per hectare per year.

According to the MOEF (2009), India's forests serve as a major sink of CO_2. The annual CO_2 removals by India's forest and tree cover are enough to neutralize 11.25 per cent of India's total GHG emissions (CO_2 equivalent), based on 1994 levels. This is equivalent to offsetting 100 per cent of emissions from all energy use in the residential and transport sectors or 40 per cent of total emissions from the agriculture sector. Clearly, India's forest and tree cover is serving as a major mode of carbon mitigation for India and the world. However, as emissions of GHGs are increasing in India progressively, efforts are needed to expand the sequestration potential of India's forests to serve to offset these.

Afforestation of available common lands—that is, community lands, roadside and canalside avenue plantations, tree planting on lands owned by institutions, defence forces and industries, rehabilitation of mined-over lands and other barren areas—will increase tree cover and carbon stocks, which will expand the carbon sink.

Policy Divergences

Inter-sectoral or cross-sector policies are often in conflict when it comes to forest conservation. For example, Table 11.1 gives a list of a few policy issues in the forestry sector that conflict with other sectors' policies.

Governance Deficit

Extant laws not fully enforced

Section 26 of the Indian Forest Act, 1927, and its variants in various states could not be enforced and all its sub-sections have been violated, resulting in serious depletion of forest resources. In a very few cases, there has been prosecution, and in extremely rare cases, there have been convictions for forest offences.

Existing policies superficially implemented

The National Forest Policy, 1988, that provides guidance for the management of forests in India is a broad statement of intent and aspirations, the objectives of which would be fully achievable only in an utopian state of affairs. As that is not the case, the required actions were not taken, resources were not made available, commensurate efforts were lacking and while there has not been a crisis situation, policy goals were only partially achieved. Some of the policy principles could be enforced only following directives from the Supreme Court of India.

Table 11.1 Policy Conflicts

Subject of conflict	Forestry policies	Other sectors' policies	Impact of conflict
Grazing on forest lands	Grazing should be minimized and regulated. Grazing is illegal in many areas.	Livestock sector: Promote increases of domestic livestock; not concerned where fodder comes from	Results in forest degradation.
Fuelwood gathering	Collection of fuelwood for personal domestic use as well as for selling is at an unsustainable level.	Energy sector: Looks to provide alternative domestic energy at a low price so that people shift to it	Results in forest depletion.
Agroforestry	Promote tree growing on agriculture land of any quality.	Agriculture sector: Land should not be diverted for raising tree crops but should be used for food production.	Supply of industrial wood depends on agroforestry.
Providing forest lands to tribal people for agricultural use.	Forest land should not be diverted for agricultural use.	Tribal welfare and development sector: Give ownership of forest land to tribal communities for cultivation of agricultural crops.	Community forest rights is a complex issue and, left unresolved, leads to further depletion of forests.
Infrastructure development	Do not use forest lands for roads, irrigation dams and canals, hydropower plants, railways and transmission lines, etc.	Transport, water resources, energy and railway sectors: Use forest lands for infrastructure-development projects.	Some forest lands will have to be sacrificed. During the next 10 years, 1 million hectares are likely to be cleared.
Mining of coal, various mineral ores	Do not clear forests for mining activities.	Coal and mining sector: Clear forest land for mining.	Some forests will have to be cleared.

The major and critical areas of policy failures are:

(a) Forests could not be fully protected
(b) Productivity could not be improved
(c) There has not been any perceptible increase in forest cover
(d) The network of protected areas (PAs) was not significantly expanded
(e) Special fuelwood plantations could not be raised to supply fuelwood on a sustainable basis
(f) NTFP resources could neither be protected nor enhanced, but on the contrary, these were left out of controlled management, giving a free and unlimited access to local communities
(g) Goal of bringing one-third of the country's land area under forests remained a distant dream
(h) Massive, time-bound afforestation, reforestation and social forestry programmes could not be sustained beyond 1992
(i) The demand and supply gaps for timber and fuelwood could not bridged; there is heavy dependence on import of timber
(j) Rights and concessions did not relate well to the carrying capacity of the forests, and nor did the capacity improve
(k) Corridors between PAs could not be developed
(l) Grazing could not be regulated or controlled despite a massive JFM programme

Thus, even 30 years after the promulgation of the National Forest Policy, any significant achievement of its objectives is yet to be seen. More than two decades is a fairly long time in the history of a nation. What went wrong is the question facing us now. Was the policy utopian, unrealistic and ritualistic or was it not seriously implemented? Perhaps it was a combination of both.

Lack of commitment, motivation and political support

Insecurity of position and tenure, lack of recognition of good performance by higher authorities and absence of equity produces poorly motivated and little-respected leaders in the forest service. The leadership crisis, coupled with bureaucratic muddles, has produced a strange work environment in the public forestry sector, resulting in demoralization and loss of the urge to introduce and accept change.

The capacity of the public-sector forestry institutions is poor in terms of the strength of the workforce and their skills. Innovation and research are inadequate. Shortage of human resources and financial constraints also affect resource management. Forest officials do not like postings in fields such as survey, demarcation, working plan and research. Institutions also do not respect technical work of a high calibre. Official complacency, indifference and lack of accountability and professional integrity have taken their toll.

The government policy relating to the carbon benefits of forests is that these will be passed on to local communities. However, as most public forests are owned by the state, significant carbon sequestration is possible from these lands. The actors in the state forest agencies do not see any incentives for them to pursue and implement REDD+ involving addressing social and environmental safeguards in accordance with the Paris Agreement. They would hardly feel motivated to work to generate financial benefits for the communities through REDD+ implementation, involving substantial and serious efforts and hard work on monitoring, reporting and verification while zealously protecting CO_2 sequestration processes. For the REDD+ incentives to be assured, its implementation has to be extended over more than three decades. The results-based finance will also involve a legal commitment to an international or foreign organization, amounting to mortgaging the forest land.

Land Tenure, Settlement, Demarcation, Mutation and the Forest Rights Act (FRA)

The tenure of forest land is not clear in many localities. The forest settlement process was completed and forest boundaries were demarcated in a majority of cases under the Forest Act of 1927 and its adapted versions in the various states. However, in many areas, the settlement process could not be completed and the legal status of vast extents of land remained ambiguous. These are mostly included in the 'protected forests' category. The Scheduled Tribes and Other Traditional Forest

Dwellers (Recognition of Forest Rights) Act, 2006, reopened the question of tenure for land that was nationalized and settled in the past and is now used and cultivated by tribal and other forest dwellers, without them enjoying any heritable and inalienable ownership of these lands. The issue of community forest rights is yet to be settled under this Act.

REDD+ Implementation in India

State-level implementation

Implementation of REDD+ activities has to be at the state level. It cannot be implemented simultaneously in all states, but must proceed in a phased manner and on a voluntary basis. Such activities will be project- and fund-driven as the states lack the resources to initiate a new activity. Forest reference and emissions levels and the monitoring, reporting and verification involve a considerable cost and also dedicated personnel, both of which are scarce. Outsourcing these activities, though it would involve higher costs, is the best alternative. REDD plus activities will also need to be funded and must provide positive outcomes in terms of reduced emissions from the forests and enhanced carbon stocks. It is possible for REDD+ to be piggybacked on an externally funded forestry-development project in a state where activities that enhance forest carbon stocks and ensure monitoring, reporting and verification may be funded by the latter. However, REDD readiness itself will require effort, personnel and finance. A number of important issues the sector is faced with will also require addressing.

REDD readiness

India does not appear to be prepared to implement REDD as of today, as it has to implement all elements and all three phases of this policy together. Preparations have not been started by the GOI, nor is India a member or partner of UN-REDD or the Forest Carbon Partnership Facility (FCPF)—two international bodies which are funding REDD readiness in developing countries. There does not appear to be enthusiasm or motivation among government agencies for REDD+; rather, there is a significant degree of scepticism.

So far a national strategy or action plan has not been developed. The drivers of deforestation and degradation have not been systematically identified. The issues of governance, gender and land tenure that need to be analysed and taken into account while developing and implementing a national REDD+ strategy or action plan have not been examined and addressed. The activities to implement REDD+ have not been identified. At the sub-national or state level too, the process of readiness has not been initiated. Since all public forests are owned and managed by state governments, capacity building and institutional development are required at that level. However, the national initiative is lacking and the states are directionless. They look to the central government for guidance in this new initiative, which is an outcome of an international agreement, and the intricacies involved in REDD+ can only be comprehended together by both the governments. Added to this is a huge responsibility for developing a forest emissions reference level and forest reference level (baseline), for which periodic monitoring and reporting would be necessary. For this, resources both in terms of skilled staff and finances would need to be mobilized.

REDD+ and the Paris Agreement

Article 5 of the Paris Agreement on global climate change stipulates that all parties to the Agreement (a) should take action to conserve and enhance sinks and reservoirs of GHGs including forests and (b) are encouraged to take action to implement and support, including through results-based payments, the existing framework as set out in the related guidance and decisions already agreed under the UNFCCC for policy approaches and positive incentives for activities relating to reducing emissions from deforestation and forest degradation (REDD); the role of conservation, sustainable management of forests and enhancement of forest carbon stocks (+) in developing countries; and alternative policy approaches, such as joint mitigation and adaptation approaches for the integral and sustainable management of forests, while reaffirming the importance of incentivizing, as appropriate, non-carbon benefits associated with such approaches.

However, it is a fact that the implementation of REDD+ is not binding if one goes by the language of the Paris Agreement as contained in Article 5. The words used are 'should' and 'are *encouraged*' (emphasis author's) and not 'shall'. This probably was a result of ecological, political and socio-economic complexities associated with the forestry sector in practically all developing countries and the enormous difficulties involved in the implementation of REDD+. Also, a consensus on REDD+ would be improbable in view of a variety of stakeholders and interests. REDD+ policies and approaches, as evolved, make it a voluntary and incentive-based activity. Nevertheless, there are international expectations that developing countries should take action to reduce carbon emissions from deforestation and forest degradation and that they should manage forests sustainably to conserve as well as enhance carbon stocks. According to Fourth Assessment Report of the IPCC, deforestation and forest degradation contribute to about 17 per cent of GHG emissions. By implementing REDD+, not only will the GHG emissions from forests be reduced but sequestration of GHG from the atmosphere will also increase.

Although REDD+ implementation is not legally binding and obligatory under the Paris Agreement, for the developing countries that have submitted their intended nationally determined contributions (INDCs), with measureable ambitions, including actions to increase forest carbon sinks, reduce deforestation (forest clearance) and degradation and undertake AR, climate-change mitigation through forests becomes binding with or without explicitly implementing REDD+ for results-based payments. Therefore, the voluntary nature of Article 5 does not remain voluntary if a country's INDCs include mitigation through forests. The implementation of INDCs is required to be quantifiable, measurable and verifiable through a system established, methodology adopted and rules framed by the COP to the UNFCCC, serving as the meeting of the parties to the Paris Agreement (the CMA, for short). This will require monitoring of carbon stocks in the forests of each country on a regular basis, with a report submitted as part of the progress report on the INDCs. For example, India has committed through its INDC to the creation of additional sinks in its forests for 2.5–3 billion tonnes of CO_2-equivalent by 2030. This may be presumed to be calculated from a baseline of 2015 levels, as the current carbon stocks in its forests must sequester and store an additional 2.5–3 billion tonnes of carbon.

A country having committed to its climate-change mitigation and adaptation ambitions (or targets) through its INDC is required to prepare a baseline or forest emissions reference level (FERL) and forest reference level (FRL) with full transparency, using the guidance and methodology of the IPCC as adopted by the COP. A periodic progress report of the INDCs will be submitted every five years to the UNFCCC secretariat for inclusion in a registry. The required information and reports submitted by a party to the Paris Agreement will be scrutinized by a technical review committee; as well, there will be a periodic stocktaking by the CMA. The provision for accountability is amply clear in the processes, methods and regulations to be developed and adopted by the CMA under the Agreement.

A country's INDC submission in itself is not an instrument for incentives in the form of results-based payments. It is an international obligation, for which the concerned country will be accountable. Unless REDD+ is implemented to achieve forest-related ambitions or targets and the results-based actions are fully measurable, verifiable and reported, no financial benefits will accrue to those countries that implement forest-based mitigation plans.

This brings us to the decision that REDD+ must be implemented in all its phases, beginning with the development of a national strategy or action plan as well as capacity. The three phases as decided by the COP are:

(a) Development of national strategies or action plans, policies and measures, and capacity building
(b) followed by the implementation of national policies and measures and national strategies or action plans that could involve:
    (i) further capacity building
    (ii) technology development and transfer
    (iii) results-based demonstration activities
(c) the above evolving into results-based actions that should be fully measured, reported and verified.

It was also decided that while developing or implementing its national strategies and national action plans, a country will address, inter alia:

(a) The drivers of deforestation and degradation
(b) Land tenure issues
(c) Forest governance issues
(d) Gender considerations
(e) Safeguards identified and agreed by the COP
(f) Full and effective participation of relevant stakeholders, inter alia, indigenous people and local communities

If a country implementing REDD+ seeks or looks for results-based finance, it is required to adhere to all the principles and take all the actions agreed upon at COP 16 (Cancun) and reiterated, finalized and adopted as part of the Warsaw Framework for REDD+ at the COP 19 session in November 2014. The sole basis for the success or failure of REDD+ is the increase or decrease in forest carbon stocks compared to a baseline or the FRL it has defined. If there is an increase, incentives will flow from the relevant sources and the increase has to be retained and desirably multiplied over the years or decades to qualify for continuing payments for these results-based actions.

REDD+ objectives can also be achieved without implementing it as an independent action for the incentives, but as part of the ambitions or targets outlined in a country's INDC to mitigate the impact of climate change by reducing emissions from forests and increasing sequestration of atmospheric carbon dioxide by expanding forest carbon sinks. Many developing countries would want to draw payments for REDD+ actions. They are in effect sequestering emissions pushed out by developed countries. For this purpose, they would need support from developed countries in the form of financial assistance, capacity building, technology and associated transaction costs, either fully or partially.

However, certain risks are associated with the implementation of REDD+. The first risk is that after substantial efforts and investment, the benefits may not be commensurate with the cost or there may even be a net negative rate of returns. Some countries that are not confident about the positive incentives from REDD+ (for example, India, China and South Africa) may avoid this risk and yet put in place or strengthen their forest carbon-inventory monitoring system.

The second perceived risk is from safeguard issues, both environmental and social. A robust environmental and social assessment (ESA) is required as an essential part of REDD+ preparedness as well as in the project-development document. Since the stringent safeguard policies of the World Bank are being adopted, many countries may not be comfortable, as mitigation plans, if any, will not only involve substantial costs but will also be cumbersome. For example, if an ESA results in the finding that the interests of indigenous people or other forest-dependent local communities are adversely affected on restricting access to REDD+ forest areas, thereby jeopardizing their livelihood opportunities, or any physical displacement is involved, safeguard issues will be triggered. The safeguards are meant to ensure that REDD+ activities also do not result in any adverse environmental or social impacts on biodiversity or any drastic alteration of a natural ecosystem's structure and composition. Implementation of safeguards may reduce or offset the financial incentives and create unnecessary responsibilities.

The third risk is the possibility of deviation from the transparency framework, as established under Article 13 the Paris Agreement with a view to building mutual trust and confidence and to promoting effective implementation. The principles and guidance relating to governance, monitoring and reporting, if compromised, will deny the offending country the expected incentives and create adverse international opinion.

The REDD+ finance for results-based incentive continues to remain elusive. The possibility of a climate fund worth US$100 billion, as discussed at the November 2016 COP 22 at Marrakech, could not be resolved by the international communities. World politics is changing fast, and so are the commitments.

Probably, the Indian government is also not very enthusiastic about the implementation of REDD+ for financial incentives on account of the lack of readiness and the complexities of a forest sector that already has numerous uncertainties and risks to contend with. Yet, India has committed through its INDC to increase forest carbon sinks and implicitly to also discourage deforestation and forest degradation. For achieving its INDC target, India needs financial resources and expects that it will flow from developed countries—so the whole commitment becomes partially conditional. As forestry institutions are not creative and innovative, there will be many missed opportunities. However, there is a possibility that, sooner or later, this issue will be given a serious consideration by the Indian Government and domestic financial resources will be used to promote enhanced forest conservation and restoration of degraded forests that will address climate change mitigation and adaptation concerns.

Appendix: National Forest Policy, 1988—A Review Matrix

	Policy Stipulation (reproduced from Government of India document)	Probable Instrument	Outcomes	Constraints
Preamble	In Resolution No. 13/52/F, dated the 12 May 1952, the Government of India (GOI) in the erstwhile Ministry of Food and Agriculture enunciated a forest policy to be followed in the management of the state forests in the country. However, over the years, forests in the country have suffered serious depletion. This is attributable to relentless pressures arising from ever-increasing demand for fuelwood, fodder and timber; inadequacy of protection measures; diversion of forest	Policy implementation through appropriate legislation and with adequate resources.	It was recognized that forests have suffered serious depletion due to a number of causes, mainly increased exploitation for fuel wood, fodder and timber as well as inadequate protection of forests and diversion of forest land to non-forest uses, as identified in the preamble.	Two major drivers of degradation—indiscriminate and unregulated, excessive grazing and exploitation for fuelwood for mainly domestic energy—could not be dealt with effectively on account of lack of will, no provisions for alternative resources or incentives

(Continued)

(Continued)

	Policy Stipulation (reproduced from Government of India document)	Probable instrument	Outcomes	Constraints
	lands to non-forest uses without ensuring compensatory afforestation and essential environmental safeguards; and the tendency to look upon forests as revenue-earning resources.		However, these issues were not been addressed effectively during the 36 years of the policy duration.	to forest-dependent communities, and socio-economic compulsions and political considerations.
	The need to review the situation and to evolve, for the future, a new strategy of forest conservation has become imperative. Conservation includes preservation, maintenance, sustainable utilization, restoration and enhancement of the natural environment. It has thus become necessary to review and revise the National Forest Policy.		The Forest Policy of 1952 was reviewed and revised due to its failure. The new policy of 1988 focused on conservation but its implementation remained inadequate during the last 30 years.	
2.1 Objectives	Maintenance of environmental stability through preservation and, where necessary, restoration of the ecological balance that has been adversely disturbed by serious depletion of the forests of the country.	Forest protection, afforestation, social forestry, reforestation, wildlife preservation, and so on.	The environment continued to be degraded. The impact of intervention has not been assessed.	'Restoration of ecological balance' is a totally abstract and unattainable objective. It cannot be clearly understood.

Conserving the natural heritage of the country by preserving the remaining natural forests with their vast variety of flora and fauna, which represent the remarkable biological diversity and genetic resources of the country.	Conservation, avoiding habitat fragmentation, maintaining connectivity and corridors.	Not achieved. Forest degradation continued. Rare and threatened species have not increased in number or extent.	Causes of deforestation and degradation not adequately addressed.
Checking soil erosion and denudation in the catchment areas of rivers, lakes and reservoirs in the interest of soil and water conservation, for mitigating floods and droughts and for the retardation of siltation of reservoirs.	Afforestation, soil and moisture conservation measures.	Indicators not defined. Remained an abstract objective/unquantifiable, hence not achieved.	An aspirational objective, more a principle rather than an objective. Gigantic task requiring huge financial resources and capacity.
Checking the extension of sand dunes in the desert areas of Rajasthan and along the coastal tracts.	Sand dune stabilization and reclamation by biological/vegetative measures.	Not achieved.	Lack of financial resources and logistically difficult task.
Increasing substantially the forest/tree cover in the country through massive afforestation and social forestry programmes, especially on all denuded, degraded and unproductive lands.	Afforestation and reforestation (AR).	Tree cover increased substantially through massive AR drives. Deforestation/degradation progressed simultaneously.	Lack of financial resources and organizational capacity. Poor maintenance of plantations.

(Continued)

(Continued)

Policy Stipulation (reproduced from Government of India document)	Probable instrument	Outcomes	Constraints
Meeting the requirements for fuelwood, fodder, minor forest produce and small timber of the rural and tribal populations.	Free access and use by local communities.	Unrestricted and unregulated use of forests continued for collection of fuelwood, small timber, fodder, non-wood forest products (NWFPs) and grazing.	Demand more than supply. Plans not made to organize supplies.
Increasing the productivity of forests to meet essential national needs	Better management, technological development and massive investment.	Not achieved. Increased dependence on imported timber and other forest products.	Financial and managerial challenges.
Encouraging efficient utilization of forest produce and maximizing substitution of wood.	Reduced production, increased prices, non-availability of local wood and cost considerations.	Not quantifiable. High prices compelled efficient utilization. Wood substitution was done by market forces.	No systematic efforts made. Market forces prevailed.

	Creating a massive people's movement, with the involvement of women, for achieving these objectives and to minimize pressure on existing forests.	Enabling environment for community participation.	JFM initiative and implementation helped in building up a people's movement. However, non-JFM areas came under higher pressure.	Sustainable availability of funds in perpetuity.
2.2	The principal aim of forest policy must be to ensure environmental stability and maintenance of ecological balance, including atmospheric equilibrium, which is vital for the sustenance of all life forms, human, animal and plant. The derivation of direct economic benefits must be subordinated to this principal aim.	Forest conservation, AR and reduced harvesting of forest products.	Not achieved due to continued degradation. Direct economic benefits derived by stakeholders, disregarding this overarching goal of the policy.	Policy implementation circumvented on account of lack of will, inadequate provision of alternate resources and incentives to forest-dependent communities, socio-economic compulsions and political considerations. Indiscriminate, unregulated and excessive grazing and exploitation for fuelwood for mainly domestic energy could not be dealt with effectively without taking drastic steps.

(Continued)

(Continued)

	Policy Stipulation (reproduced from Government of India document)	Probable instrument	Outcomes	Constraints
3. Essentials of forest management	Existing forests and forest lands should be fully protected and their productivity improved. Forest and vegetal cover should be increased rapidly on hill slopes, in catchment areas of rivers, lakes and reservoirs and ocean shores, and on semi-arid and desert tracts.	Efficient forest protection and AR, along with *in situ* soil and water conservation.	Not achieved. Encroachment on forest lands and illegal tree cutting and exploitation continued.	Financial and capacity constraints.
	Diversion of good and productive agricultural lands to forestry should be discouraged in view of the need for increased food production.	Discourage agro- and farm forestry. This contradicts with other policy stipulations aiming at increasing tree cover on non-forest lands.	Not achieved and, wherever trees gave better returns, agricultural land was used for short-rotation tree crops.	Unrealistic expectations or objective. Landowners decide the land use.
	For the conservation of total biological diversity, the network of national parks, sanctuaries, biosphere reserves and other protected areas should be strengthened and extended adequately.	A network of protected areas (PAs), with corridors and connectivity.	No significant achievement. Rights and settlements in PAs could not be finally decided.	Lack of commitment, with socio-economic and political compulsions.

Provision of sufficient fodder, fuel and pasture, especially in areas adjoining forests, is necessary in order to prevent depletion of forests beyond the sustainable limit. Since fuelwood continues to be the pre-dominant source of energy in rural areas, the programme of afforestation should be intensified with special emphasis on augmenting fuelwood production to meet the requirement of the rural people.	AR for increasing fuelwood production.	Demand exceeded production/supply, resulting in degradation of forests.	Non-reorientation of management and lack of investment.
Minor forest produce provides sustenance to tribal populations and to other communities residing in and around the forests. Such produce should be protected and improved and their production enhanced with due regard to the generation of employment and income.	Conservation and increased production of NWFP resources and also stopping destructive and unsustainable harvesting.	Not achieved.	Total decentralization and free access to for collection of NWFP.

4. STRATEGY

4.1 Area under forests	The national goal should be to have a minimum of one third of the total land area of the country under forest or tree cover. In the hills and in mountainous regions, the aim should be to maintain two thirds of the area under such cover in order to prevent erosion and land degradation and to ensure the stability of the fragile ecosystems.	Afforestation of non-forest land on a mass scale.	Could not be achieved. Rather, there is net deforestation under the Forest (Conservation) Act (FCA).	Unrealistic objective.

(Continued)

(Continued)

	Policy Stipulation (reproduced from Government of India document)	Probable instrument	Outcomes	Constraints
4.2 Afforestation, social forestry and farm forestry	A massive need-based and time-bound programme of afforestation and tree planting, with particular emphasis on fuelwood and fodder development, on all degraded and denuded lands in the country, whether forest or non-forest lands, is a national imperative.	External donor funds, rural development funds and forest-sector development funds.	The rate of afforestation and reforestation declined with the withdrawal of international donor support.	Massive failure of plantations due to poor protection and inadequate post-planting maintenance as well as lack of accountability.
	It is necessary to encourage the planting of trees alongside roads, railway lines, rivers, streams, canals and other unutilized lands under state, corporate, institutional or private ownership. Green belts should be created in urban and industrial areas as well as on arid tracts. Such a programme will help to check erosion and desertification as well as improve the microclimate.	Social forestry programmes.	Social forestry programmes proved unsustainable. Farmers carried on farm forestry or agroforestry as their own economic activity. Forest extension services dried up.	Social forestry programmes gradually declined.
	Village and community lands, including those on foreshores and in the environs of tanks, not required for other productive uses, should be taken up for the development of tree crops and fodder resources.	Social forestry, afforestation and extension services.	Partially achieved but proved unsustainable, as after harvesting, new plantations were not raised.	Long-term financial and technical support was lacking.

Technical assistance and other inputs necessary for initiating such programmes should be provided by the government. The revenues generated through such programmes should belong to the panchayats where the lands are vested in them; in all other cases, such revenues should be shared with the local communities in order to provide an incentive to them. The vesting in individuals, particularly from the weaker sections (such as landless labour, small and marginal farmers, scheduled castes, tribal communities and women), of certain ownership rights over trees could be considered, subject to appropriate regulations; beneficiaries would be entitled to usufruct and would in turn be responsible for their security and maintenance.		
Land laws should be modified, wherever necessary, so as to facilitate and motivate individuals and institutions to undertake tree farming and to grow fodder plants, grasses and legumes on their own land. Wherever possible degraded lands should be made available for this purpose either on lease or on the basis of a tree-patta scheme. Such	Relaxation of tree-cutting and transit regulations; tree patta or leasing of degraded lands.	Partially achieved. JFM programmes helped the objective. Leasing of degraded land is not acceptable to a majority of stakeholders.

(Continued)

	Policy Stipulation (reproduced from Government of India document)	Probable instrument	Outcomes	Constraints
	leasing of the land should be subject to the land grants rules and land ceiling laws. Steps necessary to encourage them to do so must be taken. Appropriate regulations should govern the felling of trees on private holdings.			
4.3 Management of state forests	Schemes and projects which interfere with forests that clothe steep slopes, catchments of rivers, lakes and reservoirs, geologically unstable terrain and other such ecologically sensitive areas should be severely restricted. Tropical rainforests or moist forests, particularly in areas such as Arunachal Pradesh, Kerala and the Andaman & Nicobar Islands, should be totally safeguarded.	FCA, 1980.	Partially achieved. However, there has been gradual dilution of restrictions.	Infrastructure development and economic growth take precedence.
	No forest should be permitted to be worked without the government having approved the management plan, which should be in a prescribed format and in keeping with the National Forest Policy.	Guidelines and working plan codes.	Achieved under the Supreme Court's orders.	Lack of technical skills, institutional capacity and motivation.

The central government should issue the necessary guidelines to the state governments in this regard and monitor compliance.

In order to meet the growing needs for essential goods and services which the forests provide, it is necessary to enhance forest cover and the productivity of the forests through the application of scientific and technical inputs. Production forestry programmes, while aiming at enhancing the forest cover in the country and meeting national needs, should also be oriented to narrowing, by the turn of the century, the increasing gap between demand and supply of fuelwood. No such programme, however, should entail clear-felling of adequately stocked natural forests. Nor should exotic species be introduced, through public or private sources, unless long-term scientific trials undertaken by specialists in ecology, forestry and agriculture have established that they are suitable and have no adverse impact on native vegetation and environment.	Research, innovation, technology upgradation and a regulatory regime.	Minor overall achievement. Impact not evaluated. Exotic species providing better returns are preferred in agroforestry programmes.	Research establishments not making the desired level of contribution.

(Continued)

	Policy Stipulation (reproduced from Government of India document)	Probable instrument	Outcomes	Constraints
4.3.4 Rights and concessions	The rights and concessions, including grazing, should always remain related to the carrying capacity of forests. The capacity itself should be optimized by increased investment, silvicultural research and development of the area. Stall-feeding of cattle should be encouraged. The requirements of the community which cannot be met by the rights and concessions so determined should be met by the development of social forestry outside the reserved forests.	Implementation with the help of law enforcement.	Not achieved.	Carrying capacity not defined and measured.
	The holders of customary rights and concessions in forest areas should be motivated to identify themselves with the protection and development of the forests, from which they derive benefits. The rights and concessions from forests should primarily be for the bona fide use of the communities living within and around the forest areas, especially the tribal communities.	Awareness-raising efforts.	JFM programmes implemented, involving local communities in forest management.	It is a principle.

The life of tribal and other poor communities living within and near forests revolves around forests. The rights and concessions enjoyed by them should be fully protected. Their domestic requirements of fuelwood, fodder, minor forest produce and construction timber should be the first charge on forest produce. These and substitute materials should be made available through conveniently located depots at reasonable prices.	Productivity improvement.	Not achieved. However, the Forest Rights Act (FRA) was enacted to settle the issue of rights over land in forest villages.	Poor growing stock.
Similar consideration should be given to scheduled castes and other poor people living near forests. However, the area which such a consideration should cover would be determined by the carrying capacity of the forests.	How to achieve this objective is not specified.	Not achieved.	Poor growing stock.
Wood is in short supply. The long-term solution for meeting the existing gap lies in increasing the productivity of forests; but to relieve the existing pressure on forests for demands of railway sleepers, the construction industry (particularly in the public sector), furniture and paneling, mine-pit props, paper and paper board, etc., substitution of wood needs to be taken recourse to.	Reduction in consumption of wood and promotion of wood substitutes.	Partially achieved. Overall productivity of forests did not improve. Timber import liberalized and made on large sale. A large number of people have been shifted to LPG.	No programme developed and implemented on a significant scale.

(Continued)

(Continued)

	Policy Stipulation (reproduced from Government of India document)	Probable instrument	Outcomes	Constraints
	Similarly, on the front of domestic energy, fuelwood needs to be substituted as far as practicable with alternate sources such as bio gas, liquid petroleum gas (LPG) and solar energy. Fuel-efficient *chulhas* as a measure of conservation of fuelwood need to be popularized in rural areas.			Conservation as a priority declining and yielding to infrastructure and economic development.
4.4 Diversion of forest lands for non-forest purposes	Forest lands or lands with tree cover should not be treated merely as a resource readily available to be utilized for various projects and programmes, but as a national asset which require proper safeguarding to provide sustained benefits to the entire community. Diversion of forest lands for any non-forest purposes should be subject to the most careful examination by specialists from the standpoint of social and environmental costs and benefits. Construction of dams and reservoirs, mining, industrial development and expansion of agriculture	Better project appraisal and use of rational criteria for conversion of forest lands.	Partially achieved. The Compensatory Afforestation Fund Management and Planning Authority (CAMPA) was established and evolved to manage funds collected on account of the conversion of forest lands. However, encroachment resulted in deforestation.	

	should be consistent with the need for conservation of trees and forests. Projects which involve such diversions should at least provide, in their investment budgets, the funds for regeneration and compensatory afforestation.			
	Beneficiaries who are allowed to undertake mining and quarrying on forest lands and on lands covered by trees should be required to repair and revegetate the area in accordance with established forestry practices. No mining lease should be granted to any party, private or public, without a proper mine-management plan appraised from the environmental angle and enforced by adequate machinery.	The conditions for approval of use of forest lands for non-forest purpose.	Not properly monitored and enforced.	Poor enforcement of conditions.
Wildlife conservation	Forest management should take special care of the needs of wildlife conservation, and forest management plans should include prescriptions for this purpose. It is especially essential to provide for 'corridors' linking the PAs, in order to maintain genetic continuity between artificially separated sub-sections of migrant wildlife.	PA management network to be extended and rationalized, and habitat to be managed better.	Corridors not provided. General decline in wildlife.	Socio-economic and political constraints.

(Continued)

	Policy Stipulation (reproduced from Government of India document)	Probable instrument	Outcomes	Constraints
4.6 Tribal people and forests	With regard to the symbiotic relationship between the tribal peoples and forests, a primary task for all agencies responsible for forest management, including the forest development corporations (FDC), should be to associate with the tribal peoples closely in the protection, regeneration and development of forests as well as to provide gainful employment to people living in and around the forest. While safeguarding the customary rights and interests of such people, forestry programmes should pay special attention to the following:		JFM launched. FRA enacted.	Regulatory mechanism set up under FRA.
	(a) One of the major causes for degradation of forests is illegal cutting and removal of trees by contractors and their labour. In order to put an end to this practice, contractors should be replaced by institutions such as tribal cooperatives, labour cooperatives,	Not a major cause as harvesting is mostly done by state forest departments or state forest development corporations.	Illegal cuttings continue but contractors have been mostly eliminated. Not achieved. Unsustainable and	Lack of strict law enforcement due to inadequate institutional capacity and infrastructure.

government corporations, and so on, as early as possible		destructive harvesting depleted NTFP.	Overlaps with the work of the tribal welfare departments.
(b) Protection, regeneration and optimum collection of minor forest produce along with institutional arrangements for the marketing of such produce	Various schemes and projects.	FRA settlements have helped.	
(c) Development of forest villages on par with revenue villages	New schemes and financial resources.	Achieved through FRA 2006.	
(d) Family-oriented schemes for improving the status of the tribal beneficiaries		Programmes implemented along with JFM for alternative livelihoods and self-help groups, capacity building of and assistance to the tribal beneficiaries.	
(e) Undertaking integrated area-development programmes to meet the needs of the tribal economies in and around the forest areas, including the provision of alternative sources of domestic energy on a subsidized basis, to reduce pressure on the existing forest areas.			
4.7 Shifting cultivation	Shifting cultivation is affecting the environment and productivity of land adversely. Alternative avenues of income, suitably harmonized with the right land-use practices, should be devised to discourage shifting cultivation. Efforts should be made to contain such cultivation within the area already affected, by propagating improved agricultural practices. Areas already damaged by such cultivation should be rehabilitated through social forestry and energy plantations.	Not achieved.	Age-old customary rights and practices difficult to change.

(Continued)

(Continued)

	Policy Stipulation (reproduced from Government of India document)	Probable instrument	Outcomes	Constraints
4.8 Damage to forests from encroach-ments, fires and grazing	Encroachment on forest lands has been on the increase. This trend has to be arrested and effective action taken to prevent its continuance. There, should be no regulari-zation of existing encroachments.		No significant achieve-ment. Efforts made to prevent encroachment.	
	The incidence of forest fires in the country is high. Standing trees and fodder are destroyed on a large scale and natural regeneration annihilated by such fires. Special precautions should be taken during the fire season. Improved and modern management practices should be adopted to deal with forest fires.		Not achieved.	
	Grazing in forest areas should be regulated with the involvement of the community. Special conservation areas, young planta-tions and regeneration areas should be fully protected. Grazing and browsing in forest areas need to be controlled. Adequate grazing fees should be levied to discourage people in forest areas from maintaining large herds of non-essential livestock.		Not achieved.	Politically not acceptable.

4.9 Forest-based Industries	As far as possible, a forest-based industry should raise the raw material needed for meeting its own requirements, preferably by establishment of a direct relationship between the factory and the individuals who can grow the raw material by supporting the individuals with inputs including credit, constant technical advice and harvesting and transport services.	Captive plantations by the industry, provision of lease of forest land and financial incentives to private sector.	Achieved to a large extent. Agroforestry promoted to meet demand for raw material for forest-based industry.	No major constraints experienced.
	No forest-based enterprise, except that at the village or cottage level, should be permitted in the future unless it has been first cleared after a careful scrutiny with regard to assured availability of raw material. In any case, the fuel, fodder and timber requirements of the local population should not be sacrificed for this purpose.	Legislation.	Scrutiny improved. Under the Supreme Court's order, the power of state governments was restricted and a Supreme Court-empowered committee assumed the power of approval. However, on the other hand, large number of saw mills, veneer and plywood mills were allowed.	Comprehensive legislation not enacted. There could be probability of subjective assessment used in some cases by the state government agencies for allowing new forest-based enterprises.
	Forest-based industries must not only provide employment to local people on priority but also involve them fully in raising trees and raw material.	No enforcement but only an advisory statement.	Not monitored. Farmers practicing agroforestry supply raw material.	Enforcement mechanism.

(Continued)

(Continued)

Policy Stipulation (reproduced from Government of India document)	Probable instrument	Outcomes	Constraints
Natural forests serve as a gene pool resource and help to maintain ecological balance. Such forests will not, therefore, be made available to industries for undertaking plantation and for any other activities.	Legislation.	Achieved to a great extent. The industrial lobby continues to make out a case for leasing of forest land for captive plantation, whenever mobilization of private investment in forestry is required.	Inadequate institutional capacity to enforce conservation policies and laws.
Farmers, particularly small and marginal farmers, would be encouraged to grow, on marginal/degraded lands available with them, wood species required for industries. These may also be grown along with fuel and fodder species on community lands not required for pasture purposes, and by forest department/corporations on degraded forests, not earmarked for natural regeneration.	Agroforestry.	It contradicts the policy objective of discouraging tree planting on good agricultural land. (Please see paragraph 3 of the National Forest Policy, 1988.)	There is no clearly defined law as the FCA, 1980 allows prior approval leasing of forest land.
The practice of supply of forest produce to industry at concessional prices should cease. Industry should be encouraged to use alternative raw materials. Import	Gradual phasing out already done.	So far this strategy has been successfully implemented.	No constraints.

	of wood and wood products should be liberalized. The above considerations will, however, be subject to the current policy relating to land ceiling and land-use laws.	Not applicable (NA).	Import of timber and wood pulp liberalized. NA.	NA.
4.10 Forest extension	Forest conservation programmes cannot succeed without the willing support and cooperation of the people. It is essential, therefore, to inculcate in the people, a direct interest in forests and their development and conservation, and to make them conscious of the value of trees, wildlife and nature in general. This can be achieved through the involvement of educational institutions, right from the primary stage. Farmers and interested people should be provided opportunities through institutions such as the Krishi Vigyan Kendras and trainers' training centres to learn agri-silvicultural and silvicultural techniques to ensure optimum use of their land and water resources. Short-term extension courses and lectures should be organized in order to educate farmers. For this purpose, it is essential that suitable programmes are propagated through mass media, audio-visual aids and the extension machinery.	Awareness raising, extension programmes.	Programmes implemented with varying degrees of success, but not monitored and evaluated to find out outcomes.	Lack of finance and institutional capacity.

(Continued)

	Policy Stipulation (reproduced from Government of India document)	Probable instrument	Outcomes	Constraints
4.11 Forestry education	Forestry should be recognized both as a scientific discipline as well as a profession. Agriculture universities and institutions dedicated to the development of forestry education should formulate curricula and courses for imparting academic education and promoting postgraduate research and professional excellence, keeping in view the workforce needs of the country. Academic and professional qualifications in forestry should be kept in view for recruitment to the Indian Forest Service (IFS) and the state forest services. Specialized and orientation courses for developing better management skills through in-service training need to be encouraged, taking into account the latest developments in forestry and related disciplines.	Government notification and support.	Forestry courses started by a number of agriculture universities. Forestry graduates inducted in the forest services through competitive processes.	Private forestry sector almost non-existent, discouraging youth from taking forestry courses in universities due to scarcity of opportunities.
4.12 Forestry research	With increasing recognition of the importance of forests for environmental health, energy and employment, emphasis must	Government support and finance.	Research establishment strengthened, with better infrastructure	Government providing inadequate finance for research. Lack of

be laid on scientific forestry research, necessitating adequate strengthening of the research base as well as new priorities for action. Some broad priority areas of research and development needing special attention are:

(a) Increasing the productivity of wood and other forest produce per unit of area per unit time by the application of modern scientific and technological methods.

(b) Re-vegetation of barren, marginal, waste or mined lands and watershed areas.

(c) Effective conservation and management of existing forest resources (mainly natural forest ecosystems).

(d) Research related to social forestry for rural and tribal development.

(e) Development of substitutes to replace wood and wood products.

(f) Research related to wildlife and management of national parks and sanctuaries.

and research priorities set up with the World Bank's (WB) assistance. However, the research programmes have not had any perceptible impact on forest productivity, reforestation, conservation, and so on.

required scientific skills and motivation. Institutional indolence.

(Continued)

(Continued)

	Policy Stipulation (reproduced from Government of India document)	Probable instrument	Outcomes	Constraints
4.13 Personnel management	Government policies in personnel management for professional foresters and forest scientists should aim at enhancing their professional competence and status and at attracting and retaining qualified and motivated personnel, particularly in view of the arduous nature of the duties they have to perform, often in remote and inhospitable places.	Ongoing action by the government.	Capacity and capability remain weak.	Low self-esteem and morale. Lack of motivation due to poor human resource management.
4.14 Forest survey and database	Inadequacy of data regarding forest resources is a matter of concern because this creates a false sense of complacency. Priority needs to be accorded to completing the survey of forest resources in the country on scientific lines and to updating information. For this purpose, periodical collection, collation and publication of reliable data on relevant aspects of forest management need to be improved, with recourse to modern technology and equipment.	Central and state governments can set up surveying and monitoring units.	FSI doing a good job of surveying and monitoring of nationwide forest and tree cover. States performed poorly on setting up and staffing resource-monitoring systems and units.	Lack of interest in the states, leading to states not developing their own establishments to survey and monitor forest resources.

4.15 Legal support and infrastructure development	Appropriate legislation should be undertaken, supported by adequate infrastructure at the centre and state levels, in order to implement the policy effectively.	Legislation.	FCA and Wildlife Act were amended; FRA enacted. However, laws not enacted to implement forest policy.	Lack of determination and will to introduce changes.
4.16 Financial support for forestry	The objectives of this revised policy cannot be achieved without the investment of financial and other resources on a substantial scale. Such investment is indeed fully justified, considering the contribution of the forests to maintaining essential ecological processes and life-support systems and in preserving genetic diversity. Forests should not be looked upon as a source of revenue. Forests are a renewable natural resource. They are a national asset to be protected and enhanced for the well-being of the people and the nation.	Enhanced financial allocation in the budgets of the states and the GOI.	Inadequate resources made available. Additional funds only became available under the FCA on account of payments for deforestation.	Lack of political support for environment and forest conservation. Low priority given to forest conservation and development.

Bibliography

Arnold, J. E. M., and Manuel Ruiz Perez. 1998. 'The Role of Non-timber Forest Products in Conservation and Development'. In *Incomes from Forests*, edited by E. Wollenberg, and A. Ingles, pp. 17–39. Bogor: CIFOR.

Bhargav, Praveen. 2011. 'Some Insight into the Forest Rights Act (FRA)'. Conservation India, 1 April 2011, http://www.conservationindia.org/resources/facts-about-the-forests-rights-act (accessed on 8 October 2018).

Bhattacharya, Pradutya, Pradhan Lolota and Yadav Ganesh. 2009. Joint Forest Management in India: Resource Conservation and Recycling. *Resources, Conservation and Recycling*, 54, No. 2010: 469–480.

Bhojvaid, P. P., M. P. Singh, Jawaid Ashraf and S. R. Reddy. 2013. *Transition to Sustainable Forest Management and Rehabilitation in Asia-Pacific Region*. Dehradun: Forest Research Institute.

B. R., Rohith. 2013. 'India Losing 135 Hectares Forest Daily: RTI.' *Times of India*, 11 June 2013. Available at https://timesofindia.indiatimes.com/home/environment/developmental-issues/India-losing-135-hectares-forest-daily-RTI/articleshow/20531915.cms (accessed on 15 November 2018).

Chopra, Kanchan, B. B. Bhattacharya and Kumar Pushpam. 2001. 'Contribution of Forestry Sector to GDP'. Unpublished report for Ministry of Environment and Forests (MOEF), Government of India, New Delhi.

Dutta, Ritwick. 2007. *Commentaries on Wildlife Law: Cases, Statutes and Notifications*. New Delhi: Wildlife Trust of India, New Delhi.

Food and Agriculture Organization (FAO). 2011. *Framework for Assessing and Monitoring Forest Governance*. Rome: FAO.

Forest Survey of India (FSI). 1999. *State of Forest Report*. Dehradun: FSI.

———. 2001. *State of Forest Report*. Dehradun: FSI.

———. 2003. *State of Forest Report*. Dehradun: FSI.

———. 2005. *State of Forest Report*. Dehradun: FSI.

———. 2007. *State of Forest Report*. Dehradun: FSI.

———. 2009. *State of Forest Report*. Dehradun: FSI.

———. 2011. *State of Forest Report*. Dehradun: FSI.

———. 2013. *Carbon Stock in India's Forests*. Dehradun: FSI.

———. 2014. *India State of Forest Report 2013*. Dehradun: FSI and MOEF.

———. 2017. *India State of Forest Report 2017*. Dehradun: FSI and Ministry of Environment, Forests and Climate Change (MOEFCC).

Government of India (GOI). 1990. *Report of the Working Group to Review the Methodology Adopted and Database Used for Estimation of GDP in Forestry Sector*. New Delhi: Department of Statistics, Ministry of Planning, GOI.

Government of India (GOI). 1999. *National Forestry Action Programme—India.* New Delhi: MOEF, GOI.

———. 2006. *Report of the National Forest Commission.* New Delhi: MOEF, GOI.

———. 2007a. *India's Forests.* New Delhi: MOEFCC, GOI.

———. 2007b. Section II, Part 1, in *The Gazette of India*, Extraordinary, 2 January 2007. New Delhi: Ministry of Law and Justice, GOI.

Indian Council of Forestry Research and Education (ICFRE). 2009. *India's Forest and Tree Cover: Contribution as a Carbon Sink.* Dehradun: ICFRE.

———. 2010. *Forest Sector Report India 2010.* Dehradun: ICFRE.

IGS (International Growth Centre). 2013. 'The Impact of FRA 2006 on Deforestation, Tribal Welfare and Poverty'. Working paper E-35032-1NC-1, International Growth Centre, London School of Economics and Political Science (LSE). London.

IPCC (Intergovernmental Panel on Climate Change), 2000. *Summary for Policymakers: Land Use, Land-Use Change, and Forestry.* Geneva: IPCC.

———. 2007. *Climate Change 2007: Synthesis Report.* Contribution of Working Groups I, II and III to the Fourth Assessment Report of the Intergovernmental Panel on Climate Change. Geneva: IPCC.

———. 2013. 'Summary for Policymakers'. In *Climate Change: The Physical Science Basis*, (WGI). IPCC. www.ipcc.ch.

———. 2014. 'Summary for Policymakers'. In *Climate Change 2014: Mitigation of Climate Change*, (WGII). Geneva: IPCC.

Khan, Irshad A. 1987. *Wastelands Afforestation: Techniques and Systems.* New Delhi: Oxford and IBH Publishing.

———. 2015. 'Bringing Back Social Forestry'. *Economic and Political Weekly* L, No. 328: 20–23.

Khosla, P. K. (ed.). 1992. *Status of Indian Forestry.* Solan: Indian Society of Tree Scientists.

Kumar, Nalini, Naresh Saxena, Y. Alagh and K. Mitra. 2000. *Alleviating Poverty through Forest Development.* Washington DC: World Bank Operation Evaluation Department, The World Bank.

Kurian, Oommen C. 2015. Implementing the Forest Rights Act: Lack of Political Will. Oxfam India Policy Brief no. 15, November 2015. New Delhi: Oxfam India.

Larsen, Jørgen Bo. 2012. Close-to-Nature Forest Management: The Danish Approach to Sustainable Forestry. In *Sustainable Forest Management: Current Research*, edited by Julio J. Diez. IntechOpen. Available at https://www.intechopen.com/books/sustainable-forest-management-current-research/sustainable-forestry-through-close-to-nature-management

Ministry of Environment and Forests (MOEF). 1988. *National Forest Policy 1988.* New Delhi: Ministry of Environment and Forests, GOI.

———. 1999. 'Study on Forest Industry'. Unpublished report. New Delhi: MOEF, GOI.

———. 1999a. *National Forestry Action Plan, Vol. 1.* New Delhi: MOEF, GOI.

———. 1999b. *National Forestry Action Plan, Vol. 2.* New Delhi: MOEF, GOI.

———. 1999c. *National Forestry Sector Project, Vol. 3C.* New Delhi: MOEF, GOI.

Ministry of Environment and Forests (MOEF). 2009. India's Forests and Tree Cover: Contribution as Carbon Sink. New Delhi: MOEF, GOI.

———. 2010. *State of Forest Report 2010*. Dehradun: ICFRE.

———. 2011. 'National Mission for Green India'. Presentation to the Prime Minister's council on climate change, 11 February 2011, New Delhi.

———. 2014. *National Working Plan Code—2014*. New Delhi: MOEF, GOI.

———. n.d. 'National Mission for Green India'. In National Action Plan for Climate Change. New Delhi: MOEF, GOI.

Ravindranath, N. H., Nalin Srivastava, Indu K. Murthy, Sumedha Malaviya, Madhushree Munsi and Nitasha Sharma. 2012. *Current Science*, 102, No. 8, 25 April 2012.

Sarin, Madhu. 1999. *Policy Goals and JFM Practices: An Analysis of the Institutional Arrangements and Outcomes*. London: WWF—India and International Institute of Environment and Development (IIED).

Saxena, N. C. 1999. *Forest Policy in India*. New Delhi and London: WWF—India and International Institute of Environment and Development.

Shrivastava, Kumar Sambhav. 2017. 'Forest Rights Act under Scrutiny'. *Down to Earth*, 17 September 2015. Available at https://www.downtoearth.org.in/news/forest-rights-act-under-scrutiny-32957 (accessed on 8 October 2018).

United Nations Developmental Programme (UNDP). 2006. 'Compendium of Basic Terminology in Governance and Public Administration'. New York: UNDP. Available at unpan1.un.org/intradoc/groups/public/documents/un/unpan022332.pdf (accessed on 8 October 2018).

United Nations Framework Convention on Climate Change (UNFCCC). n.d. 'Warsaw Outcomes'. Available at http://unfccc.int/key_steps/warsaw_ outcomes/items/8006.php (accessed on 31 July 2018).

Upadhay, Sanjay, and Videh Upadhyay. 2002. *Handbook of Environmental Law, Vol. 1: Forest Laws, Wildlife Laws and Enviornmental Laws*. New Delhi: LexixNexis Butterworths.

World Bank (WB). 1993a.*Governance*. Washington DC: The World Bank.

———. 1993b. 'India Policies and Issues in Forest Sector Development'. Report No. 10965- IN. Washington DC: Agriculture Operations Division, World Bank.

———. 1994. *Project Completion Report*. National Social Forestry Project report 1 3698. Washington DC: World Bank.

———. 1996. *Staff Appraisal Report: India Ecodevelopment Project*. Washington DC: The World Bank.

———. 2000. *India: Alleviating Poverty Through Forest Development*. Washington DC: The World Bank.

———. 2001. *Forest Strategy*. Washington DC: The World Bank.

———. 2002. *World Development Report 2002: Building Institutions for Markets*. New York: Oxford University Press.

World Resources Initiative (WRI). 2013. 'Assessing Forest Governance'. Governance of Forest Initiative indicator framework. Washington DC: World Resource Institute.

Wunder, Sven. 2001. *World Development*, 29, No. 11, November 2001.

Index

About the Author

Irshad A. Khan is an expert on forestry and natural resources management, with more than 35 years of professional experience in the areas of forest management and environmental protection. He recently (in 2012–2013) headed a USAID-funded Forest-PLUS programme aimed at REDD+ preparedness. Previously, he spent nine years as a Senior Forestry Specialist with the World Bank, where he designed, appraised and supervised forestry and watershed projects, and performed policy and institutional capacity analysis in more than seven Indian states, while serving as team leader for the World Bank's entire forestry and watershed portfolio in India.

He joined the Indian Forest Service (IFS) in 1975 and retired from the IFS in 2011. He held the post of Principal Chief Conservator of Forests with the Government of Jammu and Kashmir, and served as Head of the State Forest Department, Head of the Wildlife Department and Chairman of the State Pollution Control Board.

He has worked as a counterpart coordinated with many international donors in India's forest sector while serving the Ministry of Environment and Forests as the Deputy Inspector General of Forests and the Head of the Externally Aided Forestry Projects Division, where he oversaw the preparation, appraisal and implementation of 20 projects.

He holds a Master of Science degree in Forestry and Its Relation to Land Management from the University of Oxford, United Kingdom, as well as the Associate of the Indian Forest College PG Diploma (AIFC) equivalent of a Master of Science degree in Forestry, Wildlife and Allied Subjects from the Indian forestry college, the Indira Gandhi National Forest Academy (IGNFA), Dehradun, India. He has written several books, papers and articles.